Time
Machines

Time Machines

Scientific Explorations in Deep Time

Peter D. Ward

COPERNICUS
AN IMPRINT OF SPRINGER-VERLAG

© 1998 Springer-Verlag New York, Inc.

Published in the United States by Copernicus,
an imprint of Springer-Verlag New York, Inc.

Copernicus
Springer-Verlag New York, Inc.
175 Fifth Avenue
New York, NY 10010

Ward, Peter Douglas, 1949–
    Time machines / Peter Ward.
        p.   cm.
    Includes bibliographical references and index.
    ISBN 0-387-98416-X (hardcover : alk. paper)
    1. Paleontology.   I. Title.
QE711.2.W37   1998
560—dc21   98-18393
                                              CIP

Manufactured in the United States of America.
Printed on acid-free paper.

9 8 7 6 5 4 3 2 1

ISBN 0-387-98416-X   SPIN 10659445

Contents

Preface   *vii*

Introduction: Sucia Island   *xi*

PART ONE: FINDING TIME   *1*

1   Fossils and the Birth of the Geological Time Scale   *3*

2   Radiometric Clocks   *33*

3   Magnetic Clocks   *45*

PART TWO: PLACE   *71*

4   Baja British Columbia   *73*

5   Ancient Environments and the Level of the Sea   *105*

PART THREE: INHABITANTS   *125*

6   The Bite of a Mosasaur   *127*

7   Virtual Ammonites   *147*

8   The Ancestry of the Nautilus   *169*

9   Of Inoceramids and Isotopes   *191*

PART FOUR: THE TIME MACHINE   205

10   Cretaceous Park   207

Afterword   223

References   225

Index   231

Preface

In modern times, science has brought the past—and so many of its creatures—back to life via intellectual inquiry, application of the scientific method, and some extraordinary technology that has recently been developed. The wonder of the process is that such a rich and vivid understanding of the deep past has been generated from such scanty evidence: broken bones, lithified shells, fossil leaves, and even simple layered rocks. The scientists who have contributed to this work have woven rich tapestries of ancient times, and their weaving, which is an adventure in itself, is the subject of this book. It is as if true time machines existed, enabling us to retreat through time's mists into the past, to examine the then-living as though living still, to visit ancient worlds and reconstruct the lives their denizens led.

The past tantalizes us; it is part of our nature to seek clues about ancient times and our origins. Yet the past is far more than just some moment in time. In our own lives, for instance, it is also place, people (and other living things), and history. Take the first day of school: the desks and posters, the windows and chalkboards, the people who left us there, the people we met. So, too, for paleontologists and archeologists is the *deep* past a convergence of time, place, inhabitants, and their history or biological interactions. To bring a dinosaur to life, you must journey back to a given time, the Mesozoic Era, the so-called Age of Dinosaurs that ended 65 million years ago. But once there, you are in a place filled with life, and you must be able to understand that ancient environment and its ancient inhabitants. To reconstruct the past, it is necessary to study all of these dimensions.

In paleontology we study the past in many ways. We use theory, computers, intricate laboratory equipment, and even thought experiments. Most commonly, however, to bring the past back to life, we use rocks. The rocks we look at are usually layered or sedimentary (as opposed, say, to lava). This is the type of rock that contains fossils, and it is *fossils* that provide the key to the hoary vaults of time.

Sedimentary rocks thus contain the best information we have for studying the past—if only we can *access* that information. Any outcropping of fossil-bearing rock holds clues to its age, and thus to the age of the fossilized remains it harbors. The same rocks may also hold clues to their place of origin, which might be very different from the place where they rest now. The surface of the earth, after all, is not static but restless, its present position just a snapshot during never-ending voyages from and to. Rocks also can yield clues to the nature of the environments where they were formed: Did the materials of which they were composed "turn to stone," or lithify, on land, in a lake or sea, or in a desert? Was the environment warmer or colder, more saline or fresher, richer or poorer in oxygen or in carbon dioxide than our earth today? Most crucial of all, these same rocks usually contain our best clues to reconstructing the history of the ancient life (and death) of living forms that were present when the rocks we are studying were taking shape.

No single scientific operation or approach extracts all this disparate information. Many scientific tools, techniques, and philosophies must be brought to bear. We might call these devices *time machines*. They vary from a rock hammer to a mass spectroscope.

Many scientists spend their lives using these time machines to resurrect the past. The results can be spectacular, fueling the fascination of popular culture with dinosaurs, for example, or they can be more mundane. There are many ambiguities, however. The past no longer exists; it is really nothing but a memory—a memory of a loved one or an earth long past, of personal happiness or sadness or of creatures long dead. And memories, as we all know, are often ephemeral. Two people who attended the same event, or lived through the same history, often have quite different memories of what transpired. Interpreting the fossil record can be like that. Different witnesses—or different time machines—often yield quite different accounts of what occurred. Thus there are usually multiple versions to choose from reconstructing the past. Deciding which represents "the truth" may be difficult. Yet ambiguity is a necessary—and not entirely unwelcome—aspect of studying the deep past. It happens; people argue; the arguments are generally resolved, and scientific progress is usually (though not always) the result. Ambiguity is thus an acceptable by-product of using time machines to study the past, and it is one of the subjects of this book. It also greatly enhances the wonder and fun of doing the types of science concerned with the study of the past.

I wrote this book to show *how* the past is reconstructed by those of us who study geology and paleontology—how scientists in these fields use specific scientific techniques and instruments to resurrect ancient worlds. Part of that process involves decisions about which of many memories dredged from the past we should give most credence to. No single time machine recreates an entire picture of the past; each is like a single color or brushstroke, by itself often meaningless. Yet when combined with others, each contributes to a comprehensible—and often beautiful—portrait of the past.

To show this process, I have profiled a very specific "past" by looking at a particular time and place and at its inhabitants. Many others would do just as well, of course. The destination of my time machine was chosen simply for its familiarity and for the beauty of its time, its inhabitants, and its story. The time interval was between 80 and 65 million years ago, near the end of the Cretaceous Period, itself the last interval of the so-called Age of Dinosaurs. The place now exists in the region around Vancouver Island, most specifically a tiny island in Washington state given the name Sucia. The inhabitants are mostly dead: ammonites, mosasaurs, clams, and snails; they also include the still-living nautilus. Their story unfolds in the pages to follow.

Finally, even though the picture we paint is composed of strokes of colors discovered by a variety of scientific methods—the so-called time machines—let us again acknowledge that the actual *painting* is always done by us humans. The data blink uncomprehendingly from the myriad machines. No life animates these results without a human touch. Interpretation of the data is the final step that breathes life into the dead—the most fascinating part of the study of the deep past.

The idea of a book describing how scientists interpret the past came from Jerry Lyons, my editor on two previous books. Over several lunches, he and I envisioned a book wherein each chapter dealt with one type of analytical technique or laboratory machine that gives us a window into the past. Radiocarbon dating, functional analysis, paleoecological and other techniques—each would become an independent essay, for each is one of the "time machines" from which this book derives its title.

The wonderful ornate creation of H. G. Wells's novel *The Time Machine* is the obvious source of that title. Yet Wells's story was about the future. Our adventure beckons us to travel in quite the opposite direction. Settle in, buckle your seat belt, and ease back the controls.

Introduction: Sucia Island

Any voyage of discovery into the past must begin at some site or dig where we find evidence of ancient times and lives. A pharaoh's tomb, a mastodon's grave, a trilobite's stratum—any trip into the past begins in a modern reality that is usually dusty and cold. We who study the past have given these places a multiplicity of names: An outcrop. A stratigraphic section. An exposure. A stratum or bed. All are names for the rocks themselves, the tombs holding so many secrets of long ago. Sites that give access to the past are common. All that we need are locales with well-exposed sedimentary or layered rock and, of course, a time machine or two—the technique or laboratory procedure that scientists use to discover the nature of an ancient time, place, inhabitant, or history. The choice is thus which site to sample, which graveyard to loot, which bodies to exhume—and which kind of time machine to engage.

I have taken the liberty of picking one such graveyard as the doorway into time that provides the structure of this book. This particular portal takes us back to the Age of Dinosaurs, but at a locality that offers evidence of sea life, rather than life on the land. It is positively *writhing* with Mesozoic-aged ghosts—from the time of the dinosaurs—and is as physically beautiful as any place I have visited on earth: Sucia, a small island just south of British Columbia, where we can operate our time machines with unfettered abandon. All we have to do is get there.

The voyage starts conventionally enough. Pack the car, fill the tank, and head north out of Seattle on I-5, past the satellite suburbs and parasitic strip malls engorging themselves on the city's flanks. Soon we pass through Everett, home of pulp mills and aircraft carriers, and then into a less populated countryside. After an hour we reach Mount Vernon, looking nothing like its more famous eastern namesake. Now we leave the interstate for a highway that shoots west across the flat, tulip-and-daffodil-laden Skagit Valley, passing through fertile acreage. The dirt of these low fields finally gives way to harder outcrops. Next come oil refineries, estuarine water and sloughs, and finally the literal littoral, the shore itself. We slowly pass through the town of Anacortes, a mile of gas stations and last-chance eateries and then travel four miles along a scenic coastal road to arrive at a ferry terminus, where the wide mouths of great, squat, green and white Washington State ferries are either engulfing or expectorating cars. They are aimed north and west, toward the curiously named San Juan Island, a rocky prominence seemingly floating on a placid inland sea.

The ferry ride is unremarkable, but the scenery is spectacular whatever the season. The ferry snakes through the rocky islands, stopping here and there, heading ever northward, at last arriving at Orcas Island, most populous of the San Juan group. Once again a road leads through bucolic countryside, twisting past farms and second- or third-growth scrub forest, with the most breathtaking views given and immediately taken away. Finally a thin lane turns off the country road, private this time, descending an unhurried slope until a long gravel beach of cobble and shingle beckons. But the real show is to the north, where many small islands sit silently like Dewey's fleet anchored in Manila Bay.

Now the choice of transportation becomes tricky and improvisational. It has to be a boat capable of crossing two miles of wind-whipped sea. A Zodiac works fine, but harder boats will do. Tide is a consideration, for the North Beach of Orcas Island experiences 15-foot tidal swings; sometimes you launch right off the drifted logs that mark the highest part of the beach, and sometimes you must drag the boat across 100 feet of gravel and mud to reach the water. Gear is stowed, a prayer and a pull and the motor coughs, starts, belches gray noxious smoke and you point her north—ever northward this journey dictates—to the last bit of subaerial United States territory bowing before the great expanse of Canada.

Kelp beds (and a few drifted logs) are our only concern on the water. Sea birds keep us company as the long, low island approaches, its stone cliffs festooned with verdant arboreal banners now looming large. Landing sites are no problem; the entire island is a State Park, and two splendid docks exist for the well-to-do yachters who call this place home during the summer season. No ferry touches here, and there is no way to arrive other than by private boat.

The mooring facilities lie at the end of a large inlet appropriately called Fossil Bay. We glide in slowly, the flat water of this protected inner bay as green as the trees straining high overhead on either side. There is no escaping the cathedral-like presence of the stern rocks here. Vertical cliffs line the bay, tan in color on one side, greenish and dark on the other; these are serious rocks drenched in history—a graveyard. The bay seems to suck away sound and demands reverence. Dawn patrols of eagles are chased by gulls. Other boaters throng this harbor in summer, but they seem ephemeral and are easily ignored, ghosts among the ghosts. A few park rangers think they run the place, yet even they leave with the warm weather. Not that this matters. On Sucia Island, it is even more suitable to travel back into time in winter than in summer; it is far more lonely and deserted then, as suits a graveyard.

The long wooden planks of the island's piers give way to a path, and the tan rocks can now be seen to be massive sandstone, etched by the curious tracings that are the hallmark of water movement—in this case *ancient* water movement, for the rocks are fossilized sand dunes, incongruously frozen into rock in the middle of their ancient migrations. Abandoned quarries line the path, where great hunks of this speckled sand-dune rock were cut and carried away to become part of so many buildings erected in circa-1900 Seattle.

The tan rocks, though they are part of our story, are not the reason we have made our long journey. Such rippled rocks left behind by ancient rivers are common on our planet's surface, even if they do date to the earliest epochs of the Age of Mammals, as these rocks on Sucia do. Through scientific means that are the subject of this book, they have been shown to be a part of ancient North America, deposited here 50 million years ago, soon after a great land mass crashed into this coastline from the southwest. This continental collision was an ancient disaster, where an offshore island sailed in, millennium by millennium, like some slow ghost ship, and finally ran aground on this part of North America, welding itself to our continent in the process. The tan rocks are the first to have formed after this ancient collision. They are like a series of Band-Aids plastered over the scar marking the collision.

Continuing our march inland, we walk across a grassy plain dotted by campgrounds. More rocky cliffs loom just ahead, these made up of the thick, dark green siltstone and shale that form the other side of the eponymously named Fossil Bay.

The first fossiliferous outcrops appear: blocky, rubbly, and dark. They are olive-green sedimentary rocks but filled with pale, golfball- to football-sized hunks of limestone called concretions. How a fossil hunter loves to see these round concretions bobbing in a sea of shale! Each formed around some entombed nucleus that, more often than not, is a fossil. The first concretion sits ripe for a whack. A test blow with a hammer—the first of the many blows of a long day—loosens the arms and tells a lithic tale; the silty matrix is hard, but the round concretions are harder still, small booby-trap grenades waiting to shoot rock fragments into the eyes of the unprepared amateur. The concretions here on Sucia can be the bearers of wonderful treasure—they often contain exquisitely preserved fossilized shells.

This place is unique in the state of Washington: It is the one locality where anyone can immediately find abundant, unmistakable fossil remains from the Mesozoic Era, the so-called Age of Dinosaurs. Not dinosaurs themselves, worse luck, for none of these mighty icons have been found on the island—yet. But other types of fossils from that era are abundant. Sucia readily yields the remains of long-dead creatures so different from anything inhabiting the oceans today that they force us to confront the reality of extinction and the enormity of geological time.

Our prime locality, Fossil Bay, faces southeast, framing the enormous peak of Mt. Baker that presides over the town of Bellingham on the distant mainland. The bay is a product of glaciers, which ran parallel to the bedding of these steeply tilted rocks. Great boulders line its shore, boulders fallen from above, and everywhere the rocks are covered with a white patina of fossils, like white paint splattered by some unruly sky painter.

The dark shale and concretions alike contain these numerous fossils. They are white, or even iridescent, and they call out like beacons. Most of the fossils come from the shells of snails and clams of types long extinct. The most common and noticeable are the inoceramids; they are large and concentrically ribbed. Also immediately visible are the knobby, triangular clam fossils called trigoniids. Both of these fossils whisper "not of your time" as they fling back sunshine with their pearly luster. Other fossils here look much like clams and snails found in our world today. But none of these are the true treasures. Scattered among the smaller and more common fossils are iridescent shells in the shape of arcane spirals or straight horns, ornamented with a panoply of ribs and delicate, flower-like sutures. These are the ammonites, prizes of the Mesozoic conchologist. From inches to a foot across, they are nestled in their rocky sarcophagi, waiting for the precision blow that sets them free. A few fossil nautiloids may be found as well, along with such rarer treasures as perfectly formed shark teeth. Lost shards from a lost world.

Sucia Island, the last parcel of United States territory jutting against the immensity of Canada, is for us on this day a destination. But Sucia is also but the *start* of a journey, both geographically and temporally. It will become clear that this island, these fossils, raise more questions than they answer. A wider view will be needed if we are to find meaningful answers to questions about the age, environments, and inhabitants of this region's past. Other sedimentary rocks must be examined and compared with those found on Sucia—rocks both slightly younger and older—and for these we must travel once more. To the north and west of Sucia lie other islands with rocks Mesozoic in origin. Like Sucia, they are packed with fossil shells, and in some cases they harbor fossil bones as well. On nearby Vancouver Island, a complete, fifty-foot skeleton of an ancient, sea-going reptile of the Plesiosaur family was recovered in the early 1990s. It was one of the largest and finest examples of these long-necked marine reptiles from the Age of Dinosaurs ever recovered. A year later, the skull of a mosasaur, another type of gigantic marine reptile from the Mesozoic Era, was found on an adjoining island—a place called Hornby Island that will loom large in our travels. All of these

The south coast of Sucia Island.

places and their creatures will become our intellectual territory, the domain we will explore with our time machines.

Along the shores of the Inland Sea between Vancouver and Victoria for 100 miles and more, a gigantic expanse of strata lies draped over islands and mountains, lining bays and rivers, ready to tell tales of the Age of Dinosaurs if only the observer can unearth the information. This suite of rocks, which includes those of Sucia and Hornby and a thousand other localities, is all part of the Nanaimo Group, strata in aggregate almost three miles thick, named after a town on Vancouver Island. These rocks merit our attention for many reasons, not least for the pristine fossils they contain and the glimpses of the past they afford.

The Nanaimo Group shale that carpets this region is a remnant seabed of great antiquity, its muddy layers stacked one upon another in interminable fashion by the passing of ancient time, each older layer buried by new sediment that in the process covered the newly dead. Each successive sedimentary bed was compacted, flattened, and harrowed by worms and urchins, finally turning to hard rock and enclosing all of the collected shells from animals living in that long-ago place. But there are other rocks as well, rocks that originally formed in long-forgotten rivers on ancient beaches, or in lakes and swamps; the Nanaimo Group is the remains of 20 million years of earth history where the records of tumultuous changes in both the environment and the life forms that inhabited it are inscribed.

What were the creatures that lived in this ancient place *really* like? Where did they live? If we could only go back to the time when the gray sand and silt now frozen into stony cement were being deposited—when the ceramic-like shards wrenched and pounded from the hard rock were not the relict of *then* but part of *now*. What would that ocean bottom and those creatures look like? The only solution would be for someone, somehow to build a time machine and actually return in time to the Age of Dinosaurs. Impossible, of course. But there *are* time machines, of a sort, that enable us to reconstruct ancient fossil communities. These time machines come from science. They are the scientific method itself, along with the many kinds of devices it has spawned: mass spectroscopes, CT scanners, magnetometers,

DNA sequencers, microscopes of every stripe, ion beam microprobes, radioisotope geochemistry labs, PCR techniques, supercomputers and Silicon Graphics Indigos, and even simple rock picks are the entryways into deep time. With our modern machines there is a way to go back to a place such as ancient Sucia Island, long before it emerged from the sea to become an island of green trees and gray cliffs with the power to beguile a voyager willing to unleash the power of science to learn about the past.

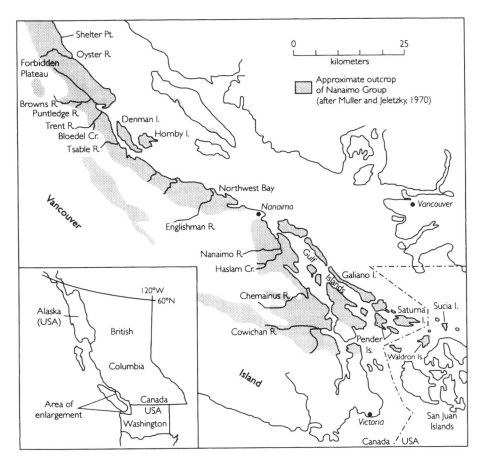

Locations of Upper Cretaceous Nanaimo Group strata in the Vancouver island region.

Many such time machines have been applied to the rocks and fossils of the Sucia Island region. Yet in that regard Sucia and its environs are far from unique. There exists an immense scientific literature detailing the myriad studies carried out in countless localities around the globe, profiling a seemingly endless series of different "pasts," each its own rich and complex world. The examples offered in this book are drawn from just one such place. The "past" of Sucia island is unique, but the way we have come to understand it is not. It is as though the shopworn science fiction concept of there being many parallel universes were indeed true, for the various time machines have revealed countless and varied "past." The past, then, seems to be made of a series of slices like those of a CT scan, where no single study reveals the whole; for an entire picture to emerge, we need many such slices, each derived from its own type of time machine.

Our first day on Sucia has passed. We sit on a gravel beach at sunset looking west from our campsite, all around us water, some real, some turned to rock. In the nearby sea, the night creatures stir. The fish and invertebrates teeming in the productive Puget Sound waters are changing shifts, and in the fossil shale we can imagine a changing of the guard as well. Ammonites rise from cold graves; fossil nautiloids resume their scavenging; snails and clams dance once more across muddy sediment. Our campfire dances too, and nearby a stir-fry on the propane stove sends delicious aroma into the sky. Venus now outshines the waning sunset, and muscles that ache from a day smashing fossils out of unwilling rock beg for attention.

Warm darkness now surrounds us like a fleece. The last light now, the fire sends sparks to rival the Summer Triangle overhead, and time slides smoothly along its two tracks—one into the deep past, when ammonites swam the sea among Mesozoic saurians, and into the near past as well, when we all first grew curious about the bestiary of the deep past. As sleep settles in, a last thought lingers about the tasks to be faced on the coming day: the study of time itself, the first step in recreating a past.

Part One

Finding Time

1

Fossils and the Birth of the Geological Time Scale

How old are the rocks on Sucia Island? How many years ago were the greenish sandstone and silts so rich in fossils *not* the stony outcrops we see today, but rather a vibrant, living seabed, home to an entire bestiary of extinct organisms? How long ago did the long-necked, scaly sea reptiles and archaic shelled mollusks live their lives? How long ago? We cannot bring back to life this world gone by until we know its age. Finding the age of Sucia Island is our first task in this journey, just as it is usually the first scientific project undertaken as by every neophyte geologist.

Geologists face the same central question in all of their studies: How old is the rock they are studying? For that matter, how old is the earth? When did a given mountain chain form, or an ancient sea? When did a particularly interesting or important evolutionary event occur? The tools we need to take these first steps in reconstructing the past vary from the simplicity of a rock hammer to the arcane electronics of an atom-separating mass spectrograph.

How do you tell time in the present? You look at your watch or the clock on the wall. And if no clock was available? You listen to the radio or watch television and wait for some announcement of the time. But say these avenues are cut off as well. Now you must rely on far more basic and crude time keepers: sunrise and sunset. Perhaps you can estimate high noon. But the middle of the night on a cloudy evening? Remove technology, and time becomes a very different—and difficult—quantity to pin down.

Without technology, time moves from the absolute to the relative—with no recourse to an external standard, one's ability to determine time becomes a relative measure. Day follows night; midday follows morning; a grumbling stomach follows sleep. Time-*keeping* in such circumstances is very crude, just like trying to date a random fossil found on a beach. You have awakened in the middle of the night, wondering what time it is; you have found a fossil on the beach, wondering how old it is. Without some contextual information and with no technology available, you are lost. But even here the metaphor breaks down. In the middle of the night you find your watch, and the riddle is solved. But with the fossil on the beach, there is no watch that will ever help you, no machine that will ever pin down this fossil's age in anything other than great crude swaths of million-year, or even hundred-million-year intervals. No wonder the early geologists wrestled so determinedly with this problem.

More than any other field of science, geology is inexorably linked to the concepts of time. The science of geology has been described as the science of telling time—establishing dates and sequences of earth events. Sooner or later any geological investigation moves from process to history, and thus to events that occurred during the passage of some discrete time interval in the past. To a geologist, measuring time is almost a religion. In no other field of science has it been necessary to codify a time scale applicable only

4

to, and usually known only to, specialists in that field. There is no formalized biological time scale or chemical time scale, even though all processes described in these two great disciplines have temporal components. All other fields of study simply use the familiar intervals of time: seconds, minutes, hours, days, and so on. Geologists, on the other hand, talk about periods and epochs, eras and zones, stages and series—the vast subdivisions of what is known as the geological time scale. Time is clearly indispensable to and intertwined with the study of geology. Perhaps this is why geologists seem a bit obsessed by time and its measurement and have been so since the birth of their science.

Efforts to address the "time problem" in geology were historically spurred by two quite different motives. The first began as essentially a religious pursuit but later became the most pressing scientific question of the late nineteenth century. Learned men (mainly theologians) were curious about the age of the earth and how that age was related to the myriad myths of the creation. The second prod was far more prosaic. Early geologists found that they could more easily find economically valuable minerals and fuels if they could understand the structure of the earth's surface. Very quickly, they realized they needed some way to date rocks.

There has certainly been no lack of effort to answer these questions. Finding the age of the earth was a quest long entrusted to theologians, who searched not in the crustal record of the earth itself but among the sacred writings of human prophets. Their answers varied between a few thousand years and infinity. The Hindu tradition weighed in at slightly less than 2 billion years, whereas some Hebrew and Christian calculations yielded values of less than 10,000 years. But as technology advanced, and the thirst not only for religious understanding but also for metals and fossil fuels increased, measurement grew more scientific.

With the onset of the Industrial Revolution, knowing the *age* of rocks became a necessary prerequisite to finding industrial minerals, such as coal, iron, and the other materials that fueled and sustaining the great Western industrialization of the eighteenth and nineteenth centuries. It was in the mining regions where engineers, who needed a better system for organizing the chaotic piles of rock slung across the earth's surface, first grappled with scientific approaches to understanding the age of any rock—and the age of the earth. They realized that if the various rock units could be dated by their relative ages, correlations among even widely separated rocks could be established and, from this, some order recognized in the geological chaos that is the crust of the earth. But how could rocks be dated?

The pioneering European geologists first believed that identifying a rock's *type* would give them a strong clue to the age of the rock formation and that one of the most powerful clues came from the hardness of a given rock. Specific rock types were thus assumed to have formed at characteristically different times, the softest rocks having formed the most recently. This crude type of dating was first used to understand the way mountains were formed. In the mid-1700s it was thought that there were three distinct types of mountains in Europe, each formed by a different type of rock and each created at a different time. According to this theory, the oldest were the Alps, which had interior cores composed of very hard, crystalline rocks (such as granite, schist, or basalt). These mountains were called *Primitive*. Sitting on the flanks of the Primitive mountains were younger *Secondary* mountains composed of layered sedimentary rocks such as limestone, often rich with fossils and intermediate in hardness. The youngest *Tertiary* mountains were composed of softer mudstones and sandstone also rich with fossils. These formed low hills rather than true mountains. Rock type and hardness thus established mountain type, and rock type also became a proxy for age. How wonderful it would have been for scientists if indeed the earth were so simply organized! Yet how disastrous for my profession! There would have been no work for the legions of bickering geologists who, it turns out, have had to do the real work of discerning the age of rocks.

Study soon exposed the fallacy of these early notions. It was discovered that some of the very high mountains were composed of the softest sediments and that very hard volcanic rock was sometimes found in very low mountains. By the early 1800s, it was understood that rock type was of little or no help in establishing either the form or the age of a mountain and that a rock's composition, or mineral content, is virtually independent of age.

Geologists were in despair until a way was found to tell the age of at least some types of rock: through the use of fossils.

Odd things, fossils. Long known as curios, they became the most cutting-edge scientific tools in the Europe of the Industrial Revolution, for it was discovered that with fossils, the relative ages of geographically separated rock bodies could be deduced. Many of us recall being fascinated by fossils as children, but they are serious tools of paleontologists even in our day—anachronisms still useful in this age of instrumental extravagance. Some of us have never put them away with the other wonders of childhood.

The first time

Stephen Jay Gould has written of his paleontological epiphany when he first encountered the American Museum's colossal *Tyrannosaurus rex* skeleton. Mine was a more modest fossil discovery, made on a seashore that now exists in memory alone.

In Washington state the strongest weather, like the strongest light, comes from the south, the direction from which great winter storms arise. Beaches, created and maintained by weather, thus take on the personality of the directions in which they face. Southern beaches carry the largest wrack, the finest sand, the biggest boulders. Western and eastern beaches are calmer, yet still prone to violence. Northern beaches are the rarest and, like the northern light and northern sea that bathe them, the most constant.

A long gravel beach claims the northern shoreline of Orcas Island. Looking north, there is only a view of largely or utterly deserted islands, mostly Canadian. One of these is Sucia Island, which from my family's vacation cabin on the northern shore of Orcas indeed looks like a low battleship. We children, brought up on patriotic World War II war movies, sank it innumerable times with imaginary torpedoes. It never moved. Easy shot.

This particular gravel beach was a biological paradise at low tide. Early morning was our favorite time for childhood beach exploration, and we searched only for sea creatures beneath the cobbles and boulders of the intertidal. The rarest treasures were small eels living in the tide pools. Rocks were not a priority. Life was.

But one rock caught my attention, in these last years of President Eisenhower's rule, as I heaved it over to unveil the cache of scurrying beach crabs beneath. Football-sized, gray, and rounded, this particular rock was absolutely packed with clamshells. Finding clamshells was no revelation, for shells in no small number littered the beach, but there was a major difference have: This rock was filled with shells turned to stone. Although some of the more far-fetched fairy tales of childhood dealt (in simpler terms) with such issues as lithification, I was by no means so unsophisticated. This rock was clearly filled with fossil shells, and with whoops and hollers I alerted the rest of my family.

A week later, a local rockhound confirmed this find as indeed being bona fide fossil material and even put a new spin on the stony clamshells. Not only were they fossils, but they were old—Mesozoic in age, he said. *Mesozoic* was a key word for an eight-year-old; Mesozoic was the time of the dinosaurs. I had found a rock with clams in it that had sat on that beach since the time of the dinosaurs, or so I divined from all of this disparate information. Yet the beach from which it came had no rock other than gravel and was not the source of this fossil. It had come from the north, perhaps from Sucia, or might even have been carried across the International Boundary during the Ice Ages 12,000 years ago when the first Canadians were eating the last of their woolly mammoths.

The rockhound's lair was a place of wonder. He had a tumbling wheel, rock saws and polishing laps, and of course rocks of all sizes and shapes littering dusty shelves. But he was also a person of some learning, and he had the gift of patience in dealing with impatient youngsters. For my part, I had made a find. I basked in glory. I asked questions. And inevitably I asked the most common question about rocks and fossils: How do you *know* how old this rock is?

He had a very difficult time answering that question in any sort of direct manner. He could only say that the clams I had found were now extinct but had lived in the Mesozoic Era. How many years ago was *that*, I asked. Hundreds of millions, he somewhat uncertainly replied. "But how do you know? Have these rocks ever been put in some machine that tells their age?" Gone now was his confidence, and I knew he was on shaky scientific ground. He lost patience with me, and my father and I soon left.

I had discovered many small truths that day and one much larger one: Divining the age of a rock can be very difficult. Years later, I would be asked the same question innumerable times and, like the rockhound of my youth, would never be able to give a short, satisfactory explanation. There is no short explanation. There is no magic box that, when placed over some dusty relict of our planet's tumultuous and seemingly ageless past, displays some shining number. It is not that we do not have machines that give ages. It is just that such machines work only on a very small suite of rock types—and even then only on an smaller subset of those rocks that have lain relatively undisturbed for great swaths of time. And tranquility is a very rare thing on this planet's everywhere-disturbed surface.

Chance and time

Childhood fossils are usually left behind on yellowing shelves. For some of us, however, they are never left behind. I could never put them away. They were transfigured from curio to curiosity.

In 1971 I became a graduate student in paleontology. I wanted to study fossil whales, but a chance event soon steered me toward a much older past, the Mesozoic-aged life and times of ancient Sucia Island. Much of my school work in that first year involved the study of stratigraphy (the geological discipline concerned with ancient time) and anatomy. But I learned as much from constant scuba diving as I did from classes at the university. I learned about the communities of organisms living in Puget Sound, who ate whom, and how these communities of invertebrates are distributed in space and time. I watched as sedimentary beds formed under the sea. I witnessed the slow burial of dead shells and bones, whose long journey toward fossildom was just beginning. Every dive was a new exploration, and with every dive I imagined what different versions of the past and its communities may have looked like. I saw each dive as a key to unlock the past.

In February I accompanied a group of new divers on a particularly cold day. We were working off the west coast of San Juan Island, in some of the most beautiful waters in all of the Americas. The huge inland sea called Puget Sound, connected to the Straits of George in Canada, is one of the most diverse (and coldest) underwater regions on earth—diverse in terms of both the animals and plants present and the various types of terrain that can be encountered.

We waded in from the beach. The dive was to be into rocky scallop beds at 50 feet or more. We shivered at water's edge, eager to get in, for this was one of the rare days when the water was warmer than the air around us. We looked like a flock of lugubrious penguins, shiny in our dark neoprene wetsuits, shuffling awkwardly in the massive equipment. When all was ready and I had absorbed the first shock of the frigid water against my skin, I replaced snorkel with regulator mouthpiece, drew a first metallic taste of compressed air, and headed down.

We passed first over gravel, then over bedrock showing the distinctive intertidal zonations of this region (barnacles followed by a mussel zone), and finally into the subtidal, where a far greater diversity of marine life greeted us. We continued downward, moving over the festooned rocky bottom, passing through pennants of green and red kelp while armies of sea urchins waved their spiny pikes at our passing shockwaves. At last we entered the zone rich in scallops, first encountered about 30 feet down, our fins creating turbulence that sent these bizarre mollusks swimming like so many wildly clacking false teeth in a bad cartoon. The water became clearer and colder, but calmer as well, the world of air and surface and all manner of that life edging away, and I felt exhilarated. I watched, and studied, and noted the relationship between depth and the type of animals present, not knowing at the time that I was laying the foundation for understanding one of the most important of all paleontological principles: that water depth can be inferred from the type of fossils present in a sedimentary section. Using the present to understand the past—the principle of uniformitarianism—is one of the most powerful of all time machines.

All too soon it was over. Air exhausted, our small group of divers retreated again to our native land. After the dive I chatted with one of the other divers, and he asked the usual questions of a stranger, including the inevitable "What do you do?" I told him that I was a beginning graduate student in paleontology and had been assigned to study whale fossils from the Washington state seacoast. The only trouble was that despite arduous searching, I had yet to find a single whale fossil and was rapidly getting discouraged. The man told me that he was an amateur paleontologist. He had never looked for whales, but he had been collecting fossils from Sucia Island for many years and had a large collection that he would give me if I wished. Apparently, I had reached a fork in my career path—one direction leading to the study of the Cenozoic Era and its whales, the other leading back into deeper time, toward the Mesozoic Era and its ammonites, nautiloids, and dinosaurs. On that day I took the latter route.

My new friend was as good as his word. Within a week an enormous collection of fossils, all meticulously cleaned, numbered, and their collection locations noted (in short, all the hardest work done), arrived in my tiny cubicle of an office. Nearly all fossils resembled the modern-day nautilus but were far more ornate. They were fossil ammonites, extinct relatives of the chambered nautilus and the first I had ever seen apart from the poor, scrubby examples in our classroom teaching set. The ammonites from Sucia are iridescent, large, exquisite. They and their kind—I thought then—are also akin to ancient stone wrist watches. All I had to do was look in some book and identify the species of ammonites from Sucia Island, and I would find the age of the sediments where they had been collected. What could be easier?

The fossils all had numbers and could be keyed into a map. But knowing the geographic position of each fossil-collecting locality was not enough for me to come up with any sense of age. Their *stratigraphic* position was what I needed. Stratigraphy is the study of layered rocks and of the relative ages of strata. Nearly all fossils come from layered rocks, and each layer in a sedimentary succession sits upon one slightly older. Because Sucia is made up of many such layers, I needed to know not only where on the island the fossils came from but also from what sedimentary level they had been collected. Without knowing this, I was duplicating the faulty way European geologists first tried to use fossils to tell time in the late eighteenth and early nineteenth Century. For a fossil to be of use, one must know its position in strata relative to other fossils: Is it from a higher or lower stratum or from a stratum of the same level? The genius of the first geologists, in the late 1700s, was their recognition that higher meant younger, lower meant older, and fossils found at the same level were about the same age. These observations became the first law of the new science of stratigraphy and were codified into the law of superposition of fossils.

I had to place the Sucia Island fossils that had been donated in the context of the stratal bedded layers found on Sucia itself. There was only one solution. I convinced several other graduate students who were studying paleontology at my University—and my academic advisor—that a winter trip to Sucia Island was a wonderful idea, a lark, a challenge, an adventure, a scientific necessity (the story changed to fit the person whose arm was being twisted).

Blackened highways led from Seattle in the rain, the short day soon giving way to night. We spent the first night in a small cottage on Orcas Island, facing Sucia. Arising and breaking fast in predawn cold and dark, we set out for Sucia with a borrowed boat and a dubious outboard engine on a steel gray dawn, leaving the coast of Orcas Island behind in the mist; soon it was lost to view. I was a summer inhabitant of these islands, and the reality of winter here was a shock. Gone were the lazy clouds and azure sky. Gone was the placid green sea. We set out in a wild winter ocean of froth and charging herds of whitened waves, an overloaded boat filled with foolish scientific zealots in search of fossils on a faraway island, cloaked in the invincibility of youth. Halfway across the two-mile stretch of open sea, the motor quit and we literally rowed for our lives, barely keeping the boat from swamping in the frigid waters. I could see that my professor was livid at having been talked into coming on such a journey in winter, and it was only the prevailing south wind that blew us to Sucia and safety.

We arrived on the dark shores soaked and frightened, conducted repairs to ensure our return at the end of the day, and then in sodden misery set out to accomplish at least something, trudging along the gravel path in wretched, squeaking boots. We students thought we knew what to do; we had read articles and taken classes. Yet our knowledge about how actually to wrest any information from these rocks was nil; we were enthusiasts, not geologists. We were like the nascent "geologists" of the late eighteenth century who knew that fossils had to be of scientific importance but had no idea how to access the information they held. My professor took charge, quoting from the work of the ghostly forefathers of geology, intoning the odd names that were becoming increasingly familiar to us: the Englishman William Smith; the French, Alcide D'Orbigny and Georges Cuvier; the Germans, Albert Oppel and Friedrich Quenstedt. He spoke of stratal succession and fossil succession, the law of "superposition of strata," and the time units called stages and zones. But he always returned to one mantra: Faithfully sample the fossils from "measured sections." Measure the strata. Collect the fossils from measured sections. He began to guide us through this thick maze of rock. "Start by measuring the sections," he kept repeating. Our impulse was to collect the fossils immediately, which is what children do. "First, describe in your notebooks the nature of the rocks while you measure their thickness, and *then* collect the fossils, always noting their exact location both geographically and stratigraphically. Unless you know the relative positions of the fossils, they are useless." My professor had been trained by Europeans. He knew that fossils could tell time only if their relative positions in the succession of strata could be determined.

We set out to do this, long measuring tapes in hand, describing the strata as best we could in our yellow field notebooks and recording the position of the fossils as we came upon them. Ammonite, field number 12/27/71-3, identity unknown, from thick limestone layer located 25 meters above the base of the section, geographic location 200 yards north of mouth of Fossil Bay." We used compasses to learn the "attitude" of the beds, measuring the degree of tilt that the strata had undergone; we proudly urinated on the rocks as we had been bidden, watching these rivulets flow down the line of the rocks' true dip. Tricks of the trade. Shrouds of rain billowed downward in this time before Gore-Tex; the tall firs overhead seemed dismally somber, and the summer joy of fossil collection now seemed grimmer, more fitting for adults, no longer a childhood delight. Lunch was a water-logged peanut butter sandwich and a can of Coke, the sugar and caffeine a welcome lift. We worked with stolid, grim purpose, and I sympathized with the long-ago pioneers who had first concocted this methodology.

As the day wore on, a column of strata began to form in our notebooks, and a checklist of fossils from various levels in that column was recorded as well. Fossil prizes were few. Up until now I had always scoured Sucia for the fossil prizes, eschewing all save the most pristine fossils. Now every scarp was information, every scrap of long-dead shell a datum point if its identity could be ascertained. The grubby fossils were chiseled out of the strata in the fine rain, later to be labeled and wrapped. The day flew by; it was time to try to return to Orcas Island in the small boat, a voyage I dreaded. I was surprised at how little one could do in a day. I was surprised that it was no longer play but work. But we had indeed collected data, and for the first time I understood responsibility to record faithfully. "Good data are immortal" read a sign on an office wall at our school. But data could be "bad"—mislabeled fossils, poor measuring, poor recording. We had taken a small measure of Sucia's secrets, its long-held mysteries. And in so doing we had proudly applied exacting scientific methodology.

Later that night, safely back in our cabin on Orcas Island, we shared our notes, shared a welcome warm dinner, shared a sense of profession. The professor watched over the apprentices with satisfaction as they drank themselves into stupor, another geological tradition.

Later that week I extracted from my still-soggy field notebook the measured section we had made and drew a column of the strata, arranged with the oldest layers on the bottom, the youngest on top. Then I portrayed in graphical fashion the positions of the ammonites earlier collected, as well as those found on this trip. Each fossil was identified as a given species and plotted on the chart, the first appearance and the last appearance of each "taxon," or species, being noted with a bold line. Soon a series of short and long lines were arrayed beside the column of strata. Some fossils were known from many specimens, some from a few or even a single one. Clearly, some of the ammonites had been very common, whereas some were much more rare. Ten different species seemed to be present. It was a start.

Fossils and "biochronology"

A very practical Englishman named William Smith was the first to show that fossils could provide a practical system for dating rocks. Nearly two centuries before our winter visit to Sucia Island, Smith was the first to formalize the system of stratigraphic collecting that all paleontologists would ultimately use.

William Smith was an English surveyor working on the excavation of the British inland canal systems in the late eighteenth century. He was present as these great canals were cut through the local rock, and thus he observed firsthand the tremendous numbers of fossils that came from the fresh cuttings. He began keeping records about the types of fossil he saw, and he eventually realized that he was seeing the same *succession* of fossil types in different regions. Many naturalists prior to Smith had recognized that fossils from a lower succession of strata were often different from the fossils found in younger, overlying strata. But no one had noticed that the succession of fossils in strata was often the same from region to region. This stroke of genius enabled Smith to "correlate" strata in widely separate geographic regions of his country.

How did the same succession of fossils come to be deposited in strata? Smith was an engineer, not a scientist; he really didn't care how the fossils he found came to be present in the rocks. To him they were simply tools to be used. In almost every case, the fossils he used were petrified shells called ammonites.

Smith's great discovery was not immediately known beyond his working-class circle of acquaintances. Indeed, his discovery was first announced in a pub. Smith was not a "gentleman," so it was many years before his discovery was published and word of it disseminated. Smith, who had had little formal schooling, found writing difficult and searched for another way to illustrate his results. He hit upon the novel idea of publishing a map of geology, not geography. Each region of differently aged rock was accorded a different color on his map, a practice still in use today. Smith, in constructing the first *geological map*, produced one of the first time machines, for his mapping of strata became one of the earliest and most accurate ways of studying ancient time.

Smith's life's work, the first geological map of England, was published in 1815. It was a revolutionary document and, like many such advances, was little noticed for many years. Smith himself received little recognition for his discovery in his lifetime. Yet gradually word of this new tool did spread, and by the second decade of the nineteenth century, many geologists began to realize what Smith had long known: Rocks could be formed at any time in earth history, but fossils could not. Limestone could be formed at any time, but limestone containing ammonite fossils could be no younger than the Mesozoic Era.

Here, then, was the new tool. You could not determine the actual age of a rock, in the sense of age as we know it. Even with a fossil, you could never identify a rock as being so many hundreds, or thousands, or millions of years old, and indeed no societies of the time could make any accurate estimate about how old the earth and its rocks were. But you *could* quite accurately determine which rock was younger and which older, and this made it possible to understand and map the structure of the earth's surface and, in so doing, discover its underground secrets and mineral treasures.

The use of fossils as relative indicators of time was soon taken up by geologists all over the European continent. Major units of time were based on the unique fossil assemblages that were characteristic of them. These subdivisions, though originally based on actual rock bodies, became de facto units of time. For instance, an English geologist named Adam Sedgewick spent several summers in the early part of the nineteenth century studying strata found in Wales. These rocks showed the transition between lower strata devoid of fossils and overlying strata filled with fossils—the transition we now know to mark the start of the Paleozoic Era. Sedgewick named the fossiliferous rocks the Cambrian *System* and used, in his definition of this group of rocks, the characteristic fossils enclosed and found within these strata. The Cambrian *Period* was defined, then, as the block of *time* during which these strata—the Cambrian System—were deposited. We now know that the Cambrian Period started about 530 million years ago and ended about 500 million years ago. Although Sedgewick's strata are found only in a small part of Wales, we refer to all rocks on earth as belonging to the Cambrian System if it can be demonstrated, through fossil content or some other technical means, that they were formed between 530 and 500 million years ago. (The assignment of ages in years to various geological units of time is a twentieth-century advance, as we will see later in this chapter.)

Even larger-scale divisions were soon recognized—defined by mass extinction events, which are sudden global catastrophes causing major biotic turnovers and extinction. Two of these were especially dramatic. At the top of strata named the Permian System, and again at the top of a much younger group of strata known as the Cretaceous System, the vast majority of animal and plant fossils was replaced by radically different assemblages of fossil. Nowhere else in the stratigraphic record are such abrupt and all-encompassing changes in the faunas and floras found. These two wholesale turnovers in the makeup of the fossil record were of such magnitude that the Englishman John Phillips used them to subdivide the geological time scale into three large-scale blocks of time. The Paleozoic Era, or "time of old life," extended from the first appearance of skeletonized life 530 million years ago until it was ended by the gigantic extinction of 250 million years ago. The Mesozoic Era, or "time of middle life," began immediately after the great Paleozoic extinction and ended 65 million years ago with The Cenozoic Era, or "time of new life," extending from the last great mass extinction (the "K/T event") to the present day.

A hierarchical assemblage of units was thus established, each based on actual rocks and the fossils they contained. The largest-scale units were the eras defined by the mass extinctions. These eras were in turn made up of the systems and their accompanying periods, such as the Cambrian and the Jurassic. Yet these units were also quite large in scale and were surely of long

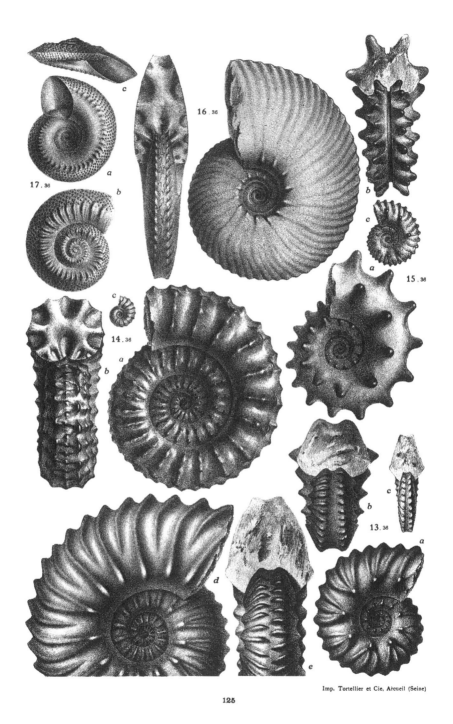

Imp. Tortellier et Cie, Arcueil (Seine)

125

Fossil ammonite cephalopods from Cretaceous Age strata, France.

18

duration. A subdivision of the periods was needed to refine geological study. There had to be a smaller-scale unit of time than the periods—a unit recognizable all over the earth. Such a unit was discovered by a French geologist working in the white chalks of Mesozoic age that are so characteristic of northern Europe, rocks ultimately named the Cretaceous System.

The great white chalk formations stretching from England across the length of northern Europe were well known to all of the early geologists, and surely to the earliest humans in Europe as well, for the chalk yielded the most important of all Paleolithic tool materials, flint. William Smith himself knew these formations well, for much of southern England is made of chalk, and his canals were cut through endless miles of the white rock. Smith recognized two units in the chalk—one greenish and sandy, the other more rich in clay. Similar units could be traced beyond England, and in 1822 D'Omalius d'Halloy completed an extensive study of the chalk in France, which he named *the Terrain Cretace,* or the Cretaceous System. Soon other workers had added to the understanding of the Cretaceous System.

French geologists discovered that many of the fossils first extracted from the chalk in England were also present in France. This discovery convinced some naturalists that the same fossils extended everywhere on earth. A French naturalist named Alcide d'Orbigny became convinced of this and, using ammonite fossils, subdivided the Cretaceous System into many smaller units. He named these *stages.* The actual basis for any stage was a fossil aggregate that was unique to it. He was sure that his units allowed the subdivision of time and that they were recognizable all over the world. Indeed, 150 years later, geologists still use both the term and the concept. The stage today is recognized as the most refined subdivision of strata—and time—that is worldwide in extent.

The various stages defined by Alcide d'Orbigny were first recognized in southwestern France. Fossils defined them, not rock type, yet by chance each had a slightly different rock type that supported a slightly different type of wine grape. The stages were named after the regions where they were first recognized, and some familiar names exist: the Coniacian Stage, named after

CIRCA 1790	CIRCA 1840		MODERN	
Post-diluvial	Alluvium		Holocene	
Diluvial	Newer Pliocene		Pleistocene	
	Older Pliocene	CENOZOIC Tertiary	Pliocene	CENOZOIC Tertiary
TERTIARY	Miocene		Miocene	
			Oligocene	
	Eocene		Eocene	
			Paleocene	
	Cretaceous	MESOZOIC	Cretaceous	MESOZOIC
	Jurassic		Jurassic	
SECONDARY				
	Triassic		Triassic	
	Permian		Permian	
	Carboniferous	PALAEOZOIC	Carboniferous	PALAEOZOIC
	Devonian		Devonian	
			Silurian	
TRANSITION	Silurian	(Protozoic)	Ordovician	
	Primordial (Cambrian)		Cambrian	
PRIMARY	AZOIC (Primary)		PRE-CAMBRIAN	

Historical development of Geological Time scale from 1790 to Modern Age.

the Conyac region; the Campanian Stage, named after the Champagne grape region. You could map the stages by the wine grapes growing in their soils. All of the great French vintages were—and are still—nurtured by the skeletons and eroded fossils of Late Cretaceous forms in France.

The first glimpse of these strata is breathtaking. The best place to view them is along a series of cliffs lining the Gironde estuary just north of Bordeaux, where the gleaming, eye-assaulting whiteness of pure chalk and limestone dominates the landscape. These strata are completely different in appearance from the drab gray and green siltstones and shale of the Vancouver Island region, yet they are rocks of the same age, visions in alabaster rather than olive.

For many years Alcide d'Orbigny described and collected from these gleaming white strata, eventually concluding that his "stages" were worldwide units. But were D'Orbigny's stages actually worldwide in extent? He certainly thought so, noting, "The stages are the expression of the divisions which nature has delineated with bold strokes." Like the great naturalist Cuvier before him, D'Orbigny was a confirmed catastrophist. He believed that the assemblages of fossils he recovered from his stages were created by some Supreme Being, lived for a short time, and then were destroyed by a worldwide catastrophe. Accordingly, they would work extremely well as worldwide time lines in sedimentary rocks.

Enthusiasm and a little knowledge (but not too much) reigned in the middle of the nineteenth century. By the 1860s the framework of the geological time scale was complete; the major subdivisions, identifying characters, and type regions had been defined. Yet so little *detailed* work had yet been done that it was optimistically assumed that assemblages of fossils first recognized in England, and later found in France and Germany, would be found the world over. Legions of European geologists streamed outward from their subcontinent, sure that the fossils now so well known in Europe would be found everywhere and could be used to assemble all sedimentary strata into what was becoming a European standard of time. But disillusionment soon set in. The farther from their own field areas the geologists ventured, the fewer familiar fossils they found. Instead they began to find new, unrecognized species intermixed with the more familiar forms of the home country. Finally, if a distant enough voyage was undertaken (such as to Africa or to the Far East), almost *no* co-occurring fossils were apparent. By the time those practicing the newly developed science of geology reached the west coast of North America, no European fossil species of any kind could be found. How could time lines be discovered without similar fossil species? The discovery of fossils in western North America in the 1860s put the methods of the European geologists to their most severe test.

Progress in mapping the vast continent of North America, especially its western portions, was excruciatingly slow. The travel alone was punishing; there were no stores nearby for provisions, no communications, no support of any kind in the event of accident or sickness. And the expenses of mapping *geology* in a new and still-poor country were prohibitive. Nevertheless, various government authorities saw the need for accurate geological maps, especially in light of the spectacular gold and mineral finds that had recently been made in California and elsewhere.

In 1860, the California state legislature created an Office of State Geology and authorized the formation of a geological survey for the state. Thus, while the Civil War raged on the other side of the continent, a small band of scientists began the immense task of completing "An Accurate and Complete Geological Survey of the State of California," which was intended to contain "a full and scientific description of its rocks, fossils, soils and minerals and of its botanical and zoological productions." Perhaps never has a geological survey been so widely defined. The new survey began its work with "greater or less vigor, according to the varying amounts appropriated by each successive Legislature." The man put in charge of paleontological study was a Mr. W. M. Gabb.

The first discovery of Cretaceous-aged rocks in California was made just east of the Bay Area, in the region of Mount Diablo, in 1861. By describing these rocks as Cretaceous, Gabb was stating that the thick mudstones and sandstone in the Mt. Diablo area were of the same age as the sepulchral chalks lining the English Channel. There could not be two more dissimilar sedimentary rock types. Gabb, however, made this correlation on the basis not of *rock* type but of *fossil* content. Both the European and the Californian rocks had one aspect in common: the presence of fossils particular to and diagnostic of the Cretaceous System as it was originally defined. These fossils were the ammonites, the group William Smith had spent so much time studying in the late 1700s and that d'Orbingy used so successfully in the mid-1800s.

Ammonites were among the fastest-evolving creatures ever to have graced the earth, and individual species are thus diagnostic of relatively short intervals of earth history. Cretaceous ammonites differ from those that came earlier by their highly ornamented shells, often uncoiled shapes, and very complicated septal sutures (the shell junctions uniting their shell walls with the chamber walls, or septa).

By 1862 the small survey crew had moved into the Great Valley of California and began surveying and excavating along the eastern flank of the mountains known as the Sierra Nevada. Deep in creek bottoms already pillaged by the gold miners of a decade earlier, they found sandstone and mudstone rich in fossils. Gabb found ammonites in abundance, the most common being a straight ammonite called *Baculites*. On two tributaries called Chico Creek and Butte Creek, these fossils were present in untold numbers, preserved in a fashion superior to any from Europe, for the dusky mudstones of California had protected the fossils from ground water and erosion. Their iridescent shells gleamed red and green as Gabb and his crew exposed them to California sunshine for the first time since the Mesozoic Era.

Gabb and his crew discovered that the Cretaceous System was widespread not only in California but along the entire Pacific Coast as far north as Canada. In 1863 Gabb took time away from his California work and traveled through Oregon and Washington to Vancouver Island, where he collected fossils in many localities. His conclusion was that many of the same species he had already encountered on Chico Creek could be found in the Vancouver Island region as well.

By the time the first Geological Survey of California had finished its initial four-year charter, it had collected thousands of fossils; and in 1864 it

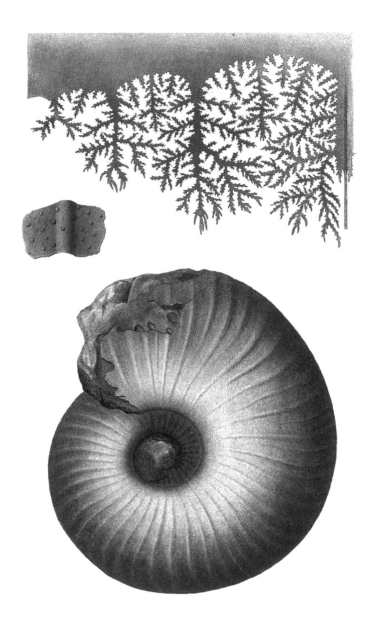

Illustration of ammonite shell from Gabb's first publication concerning the Cretaceous fossils of California.

published its first description of these fossils, a massive tome describing 260 species all new to science. Not a single one of these species was known from anywhere to the east, including Europe! The Cretaceous *System* (the large-scale time unit defined by Sedgewick and others) could be recognized, but the Cretaceous *stages* (the smaller-scale units of d'Orbigny and others), all based on ammonites from Europe, could not. These findings put an end to D'Orbigny's dream that his stages would be recognizable and useful all over the world. The Californian work of Gabb showed that only the most approximate sort of correlation could be found. More than a century later, this situation had improved only slightly. Using tiny fossil skeletons of foraminiferans, a type of free-floating amoeba with a shell, one could roughly recognize the European stages on the Pacific Coast. Many uncertainties remained when a new generation of paleontologists re-examined the rocks and fossils of California, Washington state, and western British Columbia in the middle of the twentieth century.

Time after time

The pioneering European geologists had made two great discoveries that are still useful today: (1) The most accurate way to understand the geometry of various rock bodies on the earth's surface is to make a map of them, and (2) the best way to disentangle the relative ages of various sedimentary rock units is to collect and study their enclosed fossils. The scientific study of the Sucia Island region required both.

The collections of fossils made by many paleontologists over almost a century had yielded a good idea about Sucia Island's cast of ammonites and their stratigraphic positions on the island. But the Sucia strata are but one slice of the much thicker stratal layercake found in the Vancouver Island region. Where (*stratigraphically*, not geographically) in this 15,000-foot-thick pile of sandstone, shale, and conglomerate does Sucia sit? How does its age compare with that of the sedimentary rocks found on nearby islands, such as Orcas and Waldron? How does Sucia compare in age with these and to with the more distant islands sprinkled around it? Sedimentary rock made up the entire region, but only from Sucia were fossils well described. Such was the problem facing me in 1972.

My work in this region had put me in the same boat as William Smith so long before: Smith had collected fossils from a series of isolated canals snaking across the English countryside. But which fossils—and their respective canals—were older and which younger? Had the English rocks been completely exposed (as in deserts without vegetation), he could have worked out the age relationships simply by following individual layers of rock across the countryside. But rocky exposures are the exception rather than the rule in vegetated country like England, just as they are in my island country, where the ocean creates far more geological mystery than even the trees. I had great thickness of strata, isolated by water and vegetation in many regions (around Vancouver Island), and I had a very imprecise knowledge of the age of these rocks and of how the various islands and isolated river canyons that exposed rock were correlated one to another. It was necessary to do as the early Europeans had done: spread out, map the various regions (islands and river canyons, in my case), and collect their fossils. The correct superposition of fossils, once worked out, would serve as a guide to correlation. The succession of fossils became the tool. Once I knew which ammonite followed which, I could apply this knowledge—just as Smith and D'Orbigny had done—to establishing the correct succession of rocks in the region.

The most critical regions to study are those where two separate successions of fossils are in contact—where one "zone" (based on a diagnostic assemblage of fossils) is directly overlain by another. The problem with Sucia was that no such zonal contact was exposed there; the same fossils were found in the lowest beds on Sucia as in the highest beds. Other rocks had to be searched. Sucia was a "floating" section in that it could still not be placed in the table of strata for the entire region. Other nearby islands might hold the key.

I needed a boat and so acquired my father's small aluminum fishing boat, 14 feet long with a 10-horse outboard. It was very slow. Under perfect skies I began to ply the calm green waters, using Orcas Island as a base, and radiated outward to ever more distant localities. The first stop was Waldron Island, largely uninhabited and composed of great walls of massive sedimentary rock. On my first trip there I was struck by the difference in appearance between the fossiliferous rocks of Waldron and those on nearby Sucia. The Waldron rocks were much coarser-grained. This was evidence that they had been deposited in shallower water than the rocks on Sucia. Many of the fossils were different as well. There were far more clams and snails than on Sucia and far fewer ammonites, two bits of evidence that also attested to an origin in shallow water. The ammonites that were collected seemed to belong to species different from those on Sucia. But were the differences in fossils due to differences in time or to the fact that the rock type was different? Here was another problem first discovered by the pioneering European geologists. Time was not the only variable that affected fossil content. Most animals are adapted to living in very specific environments. D'Orbigny was the first to suspect differences in fossil content could sometimes be related to his sampling different environments, not to a different time of origin. Time units had to be independent of environment, but this is rarely the case, d'Orbigny and others found. Such was my problem with Waldron Island.

Day after day I motored to this island and collected fossils from its rocky cliffs. The work was wonderful: perfect temperatures, abundant wildlife, no people. But the implications were troubling. As more and more fossils came to light, I was faced with three choices. Waldron was older, Waldron was younger, or Waldron was the same age as Sucia, but because its rocks had been deposited in a different environment, its fossils were different as a result of differences in ecology, not time. How to choose?

The answer to this riddle was supplied by William Gabb, whose work was done a century before mine. He had collected a species of ammonite from Chico Creek that turned out to be identical to those I found on Waldron Island. This ammonite, at least in California, was known to be *older* than several species of ammonites found both in California and on Sucia Island. However, several other of the ammonite species that I collected from Waldron Island were the same as those found on Sucia. Thanks to this particular correlation, I was able to propose that Waldron was slightly older than Sucia.

As the summer wore on, I studied other islands similarly. Rocks and fossils from the north shore of Orcas Island were discovered to be older than those on Waldron, and fossils on the next island in the region I studied, a place called Barnes Island, were older yet. Rocks on the first of the nearby Canadian islands to the north of Sucia turned out to be slightly younger than Sucia. By the end of the summer I had discovered the relative stratigraphic position of most islands in the region, just as William Smith had discovered the relative ages of his canals in England, and d'Orbigny the relative ages of the wine regions of southwest France. I could not assign the islands ages in millions of years. But for the first time, a table of strata could be assembled for this region. And with their relative ages, I could place the islands on a rapidly expanding geological map, the same tool—the same time machine— used by William "Strata" Smith and the other geological founding fathers.

To tell time

Biostratigraphy begins in the field where the fossils are found. It ends in a museum of some sort, for only in a museum can one make the painstaking, comparative investigations that enable one to identify fossils correctly.

The discovery and definition of the units of geological time spanned the entire nineteenth century. This revolution in the understanding and counting of geological time was propelled by huge accumulations of fossils. All this newly collected material became the bookkeeping of time, and as more fossil were collected, more and more types of fossils—types of species— poured into the various universities and geological surveys doing the collecting. All of this material had to go somewhere. The nineteenth and early twentieth centuries thus became a time of museum expansion.

The great natural history museums, such as the American Museum of Natural History, the Field Museum in Chicago, the Smithsonian Museum in Washington, D.C., and the British Museum of Natural History, were created during the nineteenth century. They are dedicated to many things, but most of all they are storehouses of species. If one finds a fossil of uncertain identity, sooner or later one must make a pilgrimage to museums where representative specimens are housed. Natural history museums are one of the most important time machines of all. It is no coincidence that they grew into stately mansions during the periods of most active geological exploration of Europe and North America.

Deciphering the age of rocks by using fossils necessitates a clear understanding of just what constitutes a fossil "species." This can be achieved only if large numbers of fossils have been collected, organized, and committed to a museum for later study. Eventually, any geologist who wants to understand ancient time using fossils ends up in a museum. In 1975 I found myself in such a place, a giant basement museum, a dark and dusty warehouse filled with metal cabinets holding corpses of ancient North America, the enormous fossil collection of the Geological Survey of Canada. Outside, the cold winter of Ottawa held Ontario in its frozen grip. Being from a warmer corner of North America, I had had no previous experience with such unremitting winter. I had traveled here to see collections assembled by all my scientific predecessors, the long string of geologists who had traveled to the Vancouver Island region to collect fossils. The dates of these entries, ranging from a century ago to only a few years before, bore witness to long-standing curiosity. Arrayed in the boxes were specimens I had seen only in monographs, fossils pristine and fragmentary, eloquent testimony to a proud geological survey's long work in deciphering the age of its northwest corner.

I had come to see more than fossils, though. Five stories above me an old Russian worked away: George Jeletzky, one of the giants of geology and a world authority on Cretaceous ammonites. Jeletzky was a legend for many reasons. He was prolific, he was a pioneer in Cretaceous geology, he was a refugee: He had fled the Soviet Union during World War II, his fossils his only possessions, and made his way to Canada. He still had a price on his head and faced execution if he ever returned to his native Russia. Warm, humorous, and earthy, he was thus exiled forever for crimes imagined or political (if there is any difference). There may be no greater punishment than separating a Russian from his country.

Jeletzky had spent years in the wilderness of western Canada, deciphering the rock record of the Cretaceous there and splicing together the threads of insight afforded by the insular Canadian fossils with those from the rest of the world. Perhaps better than anyone on earth, he was capable, using fossils, of drawing the fine lines of time that relate the various geographic regions of this far-flung earth. He had trekked over the entire region of Vancouver Island. Yet Sucia still remained an enigma. Some of its fossils were unique in the world, or so it seemed. It was still not certain how the shale of Sucia correlated not only with other Cretaceous-aged rocks in Europe but even with rocks in California and some regions of Vancouver Island.

The best fossils for determining ages for the latter parts of the Cretaceous are the curious, straight-shelled ammonites called *Baculites*, forms found abundantly in the Vancouver Island region. Although the majority of ammonites during their long, 360-million-year reign on earth were coiled like a nautilus shell, some, in the Cretaceous, tried the new tack of uncoiling their shells. The *Baculites* were one of the more extreme manifestations of this uncoiling. Their shells were long tubes, and some became as large as a human. Most, however, were small. Nearly all of the *Baculites* showed two distinctive characteristics that made them especially useful for dating rocks. First, they were extraordinarily abundant. On Sucia there are easily ten times as many *Baculites* ammonites as any other kind, and this is the pattern reported around the world. Second, for reasons still entirely unknown, the *Baculites* lineage showed very high rates of evolution and extinction. Each of their species lasted on earth but a short time—often a million years or less. Thus they are among the best of all time markers, for their presence in sedimentary strata pinpoints very short intervals of time. They can allow precise correlation between distinct rock assemblages.

The western interior of the United States has been subdivided into units of a half-million years or less, on the basis of the presence of *Baculites* species. New species of *Baculites* ammonites appeared to have been forming in the vast western interior ocean at a prodigious rate between 90 and 65 million years ago, and thanks to their amazing abundance, paleontologists for more than a century had been able to use these fossils to subdivide the thousands of feet of shale in the western states. Such, however, was not the case for the west coast of North America. Although a very good Japanese paleontologist named Tatsuro Matsumoto had recognized a handful of *Baculites* species from California in the mid 1950s, only two species had been found in the Vancouver Island region, and one of these came from Sucia Island. More disappointing yet was the discovery that unlike the case in the western interior, where *Baculites* species appeared and disappeared with great abandon, the two *Baculites* from the Vancouver Island region just seemed to go on and on and never go extinct, lasting many millions of years. It made no sense to me. Why would speciation of this type of ammonite be so abrupt in one place and so gradual in another?

My agenda in Ottawa in that cold winter was to look anew at the *Baculites* species known from the Vancouver Island region and compare them with others of the same group. Over many days I lived in the dark basement, pulling from bulk collections the numerous *Baculites* specimens collected over many years. With hundreds of fossils to look at, it quickly became clear that just as I had suspected while in the field on Vancouver Island, there *were* far more than the two species that Jeletzky had identified. I was electrified by this knowledge, for after nearly three years of work in this newfound paleontology, I had finally made a discovery that would affect the understanding of geological time and its correlation from region to region. Species well known in California—but previously unreported from Vancouver island—were clearly present in rocks slightly younger and slightly older than those in Sucia Island. With this new information I could for the first time correlate the Sucia rocks with California, Alaska, and Japan. Perhaps five other people in the world would care. But those five would really care! The soldiers in the paleontological army are few, but they are dedicated.

The next step was far more delicate, for I had to convince Jeletzky that I was correct—and that he was wrong. It had been he who had stated that only two *Baculites* species were found in the Vancouver Island region. We met in his office (he was in his sixties then), and he inspected my evidence. To my surprise he readily acquiesced to my changes in his findings. He laughed. "Let me tell you a secret," he said. "I never liked those *Baculites* fossils. They all look alike to me. I never bothered to really study the Vancouver Island species." Another torch had been passed. Teacher and student, monk and acolyte; master and apprentice at the end of a long apprenticeship—two equals now reveled in one more fact beguiled from nature's grasp. Sucia became bracketed in time that day.

The long road traveled to arrive at the relative age of this particular group of rocks is typical not just of Cretaceous rocks but also of rocks and fossil from throughout time. The scientists doing the studying and the times, rocks, and fossils they study always vary, but the methodology is remarkably similar. Any study starts with one section of rocks and then expands, eventually ending up in a museum. It is an example of nineteenth-century science still working well on the threshold of the twenty-first century.

The work of Gabb and others in North America showed that crude but reliable worldwide correlation of fossil-bearing strata was possible. My work a century later was one of many confirmations of the efficacy of this type of time machine. Sedimentary rocks could be dated by studying their fossils. However, for most other rock types—such as igneous and metamorphic rocks—assigning *any* sort of age was impossible before the twentieth century. The problem was that fossils are found only in sedimentary rocks, never in igneous or most metamorphic types. Furthermore, even in sedimentary rocks fossils are only occasionally abundant, and in many sedimentary rocks (such as glacier, river, and desert deposits) they are never present at all. Some method other than the use of fossils had to be developed for the purpose of dating rocks. Ultimately, several such systems were established. The most commonly used of these newer systems, radiometric dating, is the subject of the next chapter.

2

Radiometric Clocks

When ammonites still lived and died and fell into the sandy bottom of what is now Sucia Island, North America was a divided continent. For tens of millions of years during the Late Cretaceous Period, a giant seaway hundreds of miles wide cut the continent in two. From the arctic regions of the far north to the Gulf of Mexico in the south, this wide yet shallow sea, called the Western Interior Seaway, was home to an incredible bestiary of now-extinct creatures. Archaic fish and giant marine reptiles prowled the blue waters in search of prey; monstrous turtles sculled among the wavetops; and giant clams were so abundant that their shells paved the muddy bottoms. Overhead, long-winged pterosaurs and primitive birds circled above the whitecaps, diving and sometimes fighting for surface-dwelling fish. Ammonites lived and died in countless numbers in the sunlit portions of the sea, their empty shells littering the shallow sea bottoms and sandy shorelines. To the west of this sea the Rocky Mountains were rising, urged skyward by huge volumes of magma welling up among their roots. In places this magma pooled far underground, slowly solidifying to become giant bastions of granite, the speckled foundation of any continent. Elsewhere in the rising arc of mountains, the magma successfully fought to the surface, blasting outward from volcanic cones, creating smokestack pillars of blackness to the west of the inland sea. Falling ash from the volcanoes periodically covered the shorelines of the seaway, creating a rich, fertile soil.

Riotous jungles grew and died in the swampy lowland areas along the margins of the great sea. Rivers large and small poured into the seaway, carrying to the central sea unnumbered tons of sediment derived from the new mountains as they began to erode. Carried within or buried beneath this settling sediment were the remains of many creatures of that time: the skeletons of fish, the shells of long-extinct mollusks. Yet the most valuable product for geological dating in this region came not from the fossils but from the volcanic ash that fell on this land. The intersection of this fine volcanic dust with sedimentary rock gives us a means of dating the fossiliferous sediments in terms of absolute years from the present, not merely relative position. The ash can be dated *radiometrically*. This methodology is among the most scientifically useful of time machines.

The age of the earth

There has always been a need to put absolute age dates—the time, in thousands or millions of years ago, that given rock formed—on the units of the geological time scale. Fossils alone cannot do this. They can help us sort out relative positions of strata but cannot gives us their age in numbers of years ago. The solution to that problem awaited the discovery of radioactivity and the invention of machines that can measure the tiny quantities of chemicals called isotopes, that are used to date a rock. Those discoveries took place in the twentieth century, but they were not spurred by the need to add absolute age dates to the geological time scale. Radiometric dating, the discipline of assigning dates to rocks, developed in response to an entirely different question: finding the age of the earth.

By the late 1800s, the search for a reliable means of dating nonsedimentary rocks was a major goal of science. Economic motives were still important, just as they had been at the start of the century. But as the century came to a close, a more purely scientific reason for the better dating of rocks emerged. Scientists of the time became much interested in determining the age of the earth itself. There were many motives for this pursuit, but perhaps none so pressing as those related to Darwin's newly proposed theory of evolution. This theory necessitated an earth of great antiquity, and both Darwin's supporters and his detractors clamored for better information about the age of the earth.

Discovering the age of the earth was perhaps the most technically difficult challenge facing scientists of the late nineteenth century. Two schools of thought prevailed. Most geologists, recognizing the immense thickness of sedimentary rocks scattered over the globe, believed that *billions* of years had transpired since the origin of the earth. The great diversity of fossils encased in all these strata seemed to support this view, because Darwin's theory of organic evolution required that great expanses of time must have elapsed since the origin of the earth to allow for the evolution of organisms now present in such great diversity. By contrast, physicists thought the earth was much younger than this, perhaps a few hundreds of millions of years old. These clashes, the first to pit against each other the physical scientists and those who styled themselves "naturalists," were a harbinger of the future, and they were immensely bitter.

At this time, the man most notable for addressing the question of the age of the earth was William Thomson, better known by the title of his peerage: Baron Kelvin of Largs Ayrshire—Lord Kelvin for short. Thomson has been described as the most honored British scientist in history, and he was certainly the most famous scientist in the world at the end of the nineteenth century. He was elected president of the Royal Society and was re-elected five times. During his illustrious career, Thomson, or Kelvin, published over 600 scientific articles and books. It was he who, through brilliant insight and mathematical calculation, made the first *scientific* computation of the age of the earth. Kelvin's results seemingly struck a mortal blow to Charles Darwin's theory of evolution, for Kelvin's estimate—that the earth was created less than 100 million years ago—did not seem to allow enough time for the myriad creatures found on earth to have evolved.

Kelvin's estimate was based on his scientific passion: the study of heat and heat flow (the absolute temperature scale bears his name). It was known at that time that each transformation of energy from one form to another results in the formation of heat, which is then dissipated. This fact, which eventually become known as the second law of thermodynamics, is the reason why perpetual motion machines are impossible—energy is always lost by any machine, and in every transfer of energy. Kelvin reasoned that the earth and the sun are both cooling and that, by establishing the initial temperatures of each of these bodies, as well as their rate of heat loss, he could arrive at an estimate of their age. Measuring the rate of the sun's heat loss was difficult, but measuring that loss for the earth was not. Kelvin needed just three values: the initial temperate of the earth, the thermal conductivity of rocks, and the actual heat flow. He used a value of 3870°C as the initial temperature of the earth, because this was thought to be a reasonable estimate of the temperature at which rocks melt to a molten state, and because he assumed that the earth formed from an originally hot, magmatic body (rather than coalescing from many cold fragments). Heat flow measurements for various types of rocks were obtained in the laboratory. All that was wanting was a measure of the actual rate at which heat was escaping from the earth's interior, and this was found by measuring the earth's temperature at various depths in underground mines. Not only did these experiments indicate that heat is indeed flowing outward from the deep interior of the earth but the initial rate they suggested was so high that it implied a relatively young age for the earth.

Kelvin's calculations were considered highly authoritative, not so much for *how* they were done but because of who Lord Kelvin was. Kelvin's conclusion about a relatively young earth was soon supported by others who studied heat flow, most notably by Clarence King, director of the Geological Survey of Canada. King, using Kelvin's methodology, arrived at an even lower estimate than Kelvin's; he calculated that the earth was only 24 million years old.

The Kelvin–King conclusions regarding the age of the earth were widely but not universally accepted. The chief dissenters were those familiar with evolution and the fossil record—such as Darwin's "bulldog," Thomas Huxley—and those who studied the rate at which sedimentary rocks accumulate. Twenty-four million years was far too short a time to accommodate all of the evolutionary changes visible in the fossil record, and it was also too short to account for the amount of sedimentary rock present on the earth's surface. With regard to the latter, in 1895 the geologist William Sollas estimated that thickness to be more than 100,000 meters, if it were to be piled upward in one continuous column. A column of sediment 60 miles tall, sitting on the surface of the earth, would certainly take time to form, assuming that the processes (and rates) of sedimentary rock accumulation have been uniform through time. Sollas then asked the following question: What *is* the rate at which sedimentary rocks, past and present, can be expected to accumulate? He assumed that sedimentary rocks accumulate at a rate of about 1 meter each 300 years and thus arrived at an estimate of about 34 million years for the age of the earth since sedimentation began. This figure was very much a *minimum*, for it assumed that the sedimentary rocks accumulated without pause or break. In fact, however, most sedimentary rocks accumulate in sporadic fashion, and there may be long periods of time when no sediments accumulate. Furthermore, the Sollas estimate did not account for the long stretch of time that passed, because of the earth's early, molten surface, before sedimentation even began.

A more ingenious attempt to arrive at a reliable estimate of earth's age came to be known as the "salt clock." The amount of salt in the ocean, and the rate at which it arrived there, were thought to offer a way of calculating the age of the earth. Using this method, several chemists arrived at an estimate of about 100 million years.

By the earliest part of the twentieth century, physicists, chemists, and geologists had reached consensus on a figure of between 25 and 100 million years for the age of the earth. Only the evolutionists found this number implausible, for they believed that the workings of organic evolution required much longer stretches of time to generate all the organic diversity found on earth today. As we know now, their misgivings were well founded. The true age of the earth is measured in billions of years, not millions. This discovery came about as a by-product of developments in the field of nuclear physics, which eventually gave us a means of dating volcanic ash.

Near the close of the nineteenth century, at the peak of Kelvin's eminence, the seeds of his downfall—in terms of his estimate of the earth's age—had just been sown. In 1895 Wilhem Roentgen of German discovered X-rays, and a year later A. Becquerel of France discovered the radioactive properties of uranium. Two years after that, Marie Curie discovered similar properties for the element thorium and coined the word *radioactivity*.

Radioactive decay

The breakthrough leading from the first discovery of X-rays and radioactive elements to accurate age determination in geology came from Ernest Rutherford and Frederick Soddy of McGill University in Montreal, who discovered the principle of radioactive decay and "half-life." In radioactive decay, an atomic nucleus undergoes transformation into one or more different nuclei, the original element often becoming in the process a different element. By 1907 Rutherford made the first suggestion that radioactive decay processes could be used as a geological timekeeper:

> If the rate of production of helium from known weights of the different radio-elements were experimentally known, it would thus be possible to determine the interval required for the production of the amount of helium observed in radioactive minerals, or, in other words, to determine the age of the mineral.

These prophetic words heralded a new era of age determination. Rutherford quickly calculated the age of two minerals, finding *minimal* ages of 500 million years for each—far older than the greatest age of the earth it-self as calculated by Kelvin and others. Soon other workers began analyzing rocks that contained radioactive elements. By 1907 a slightly improved method yielded a series of ages all much greater than the 100 million years proposed by Kelvin as the maximal age of the earth. The greatest of these was 2.2 billion years. The discovery of radioactivity—and the natural clock with which it keeps time—made Kelvin's simple conductive calculation ob-solete. Kelvin's mistake was to assume that no new heat had been created in the earth since its origin. Radioactive decay produces such "new" heat.

The new method of age dating, based on radioactive decay, required ac-curate measurement of the isotopic ratios of various elements. The isotopes of an element are atoms of that element that have the same number of protons but different numbers of neutrons. Perhaps the best-known example is carbon-14, which is found in very small quantities compared to its far more common sis-ter isotope, carbon-12 (the numbers refer to the atomic weight of the atom). The transformation of one isotope into another that we call radioactive decay takes place at a known and constant rate. There are 339 isotopes of 84 ele-ments known in nature; 269 of these are stable (they do not change), and 70 are radioactive. Of the 70 radioactive isotopes 18 are too "long-lived" to be useful; they have too long a half-life (the amount of time required for exactly half of the parent isotope to be transformed into the daughter isotopes).

The early geochronologists examined only the breakdown of uranium into lead for age dating. Today other isotopic pairs are examined, the most useful being potassium/argon and argon/argon. The potassium/argon (K/Ar) method measures the relative amounts of potassium-40 and successor argon-40. This method is the most widely used of all radiometric dating method-ologies because it is cheap, reliable, and the method least likely to be com-promised by contamination. Because argon is an inert gas, it never combines chemically with any other material in the rock, and it escapes as a gas when a rock is heated. Potassium is one of the most abundant elements in rocks and is constantly breaking down into argon, which then escapes in most cases. However, under special circumstances the potassium/argon "clock" is set in motion in such a way that geologists always know that they are dating the true formation of a given rock: These special conditions occur when molten magma cools and solidifies. Any lava has a great deal of potassium, and the argon being produced by radioactive decay is not trapped because of the heat and liquidity of the lava. When the lava cools, however, crystals of feldspar and other minerals form, each containing potassium-40 that is decaying to argon-40 at a rate such that half of the potassium will be transformed into argon in 1.25 billion years. The moment the crystal forms, the clock is set at 0.

Recently, the K/Ar method has been improved. Samples with potassium crystals are irradiated with fast neutrons in an atomic reactor, which converts a fraction of potassium-39 to argon-39. The ratio of the two argons is then examined. This technique is the most precise of all dating methods, but it can be used only on rocks a million years of age or older. In younger rocks there just has not been enough time to cause measurable accumulations of the daughter isotopes.

By the earliest part of the twentieth century, much of the theory behind radiometric age dating had been worked out. What was necessary to apply the theory, however, was a reliable set of instruments. The physicists needed to measure tiny quantifies of specific isotopes with great precision. It was now up to the instrumentalists to put theory into practice.

The machine necessary to measure various isotopic quantities is known as a mass spectrograph. The first of these elegant machines was designed by the physicist J. J. Thompson of the Cavendish labs in Cambridge, England. His crude prototype was soon redesigned by a colleague in the same lab, F. W. Anston, who named his creation the "positive ray spectrograph." With this machine, Aston began measuring the isotopic compositions of many materials. In 1927 he turned his attention to the radioactive decay of lead isotopes, and in so doing launched the field of research that eventually gave us some of the most profound insights into not only the age, but also the formation, of the solar system.

Mass spectroscopes have speciated and multiplied over the years, and now they are quite task-specific. They are among the most common of major laboratory instruments; thousands are operated daily throughout the world. Some investigators continue to look at the relative isotopic composition of earth material to produce reliable dates. Others examine lighter isotopes of carbon and oxygen in order to measure ancient temperatures. Many scientists, however, are still dedicated to finding the age of rocks and to solving the most interesting question of all: How old is the earth? The most recent measurements suggest an age of at least 4.5 billion years.

Returning to rocks

Ironically, the techniques of radiometric age dating work best on those rocks that most stubbornly resisted nineteenth-century efforts to determine ages—the igneous rocks, which never contain fossils. Radiometric dating methods work least well on those rocks that contain fossils. The most sophisticated mass spectroscope in the world is useless on an ordinary fossil-bearing sedimentary rock. If a sedimentary rock is to be dated radiometrically, there must be some type of interbedded lava flow or ash layer locked within. Unfortunately, such events were relatively rare.

Just as clearly, the majority of sedimentary rocks cannot be directly dated. The only solution was to find a way to date rocks in those places where sedimentary rock units and volcanic rock units intersected in time—places where ashes or lava flows can be found interbedded with ancient sedimentary rocks. Such places became the new holy grail that geologists sought. Paleontologists and other geologists learned a new word and learned to recognize a new rock type: bentonite, a thin orange layer of lithified volcanic ash found in sedimentary rock.

Arriving at age dates for Cretaceous sedimentary rock has been a particularly vexing problem. The original type areas exposed in Europe were deposited in regions far from active volcanoes so they have few or no bentonites. Much better beds are found in the western interior of North America, east of the Rocky Mountains—the site of the ancient Western Interior Seaway.

The microscopic ash particles erupted out of the ancient Rocky Mountains during the Cretaceous Period are the source of our best age dates for the Age of Dinosaurs. When these microscopic crystals first hit the air, they cooled, which allowed tiny feldspar crystals rich in potassium to form. Most of these ash particles hit the surface of a wide inland sea east of the volcanoes, the seaway that bisected the North American continent between 100 million and about 70 million years ago.

The ash that fell on the land was quickly washed away by rain and destroyed. But, each monumental volcanic burst also sent a shower of ash onto the seaway's broad surface, where it slowly sank, finally to come to rest on deep, murky sea bottoms where fish and shells—and above all ammonite cephalopods—lived and died in numbers beyond counting. Inside each of the tiny crystals of feldspar forming this sunken ash, isotopes decayed from one to another like, well, clockwork.

The thick shale, sandstone, and ash deposits of the Western Interior Seaway can be seen throughout the central part of North America. In Colorado, Wyoming, the Dakotas, Arizona, Alberta, Manitoba, Montana, Nebraska, and Kansas, the sedimentation of more than 50 million years is visible. Better than any other place on earth, this stack of sediment contains markers, both fossil and igneous, that can be used to tell time, Cretaceous time.

The fossils alone make the region extraordinary. The best-preserved ammonites in the world come from here—beautiful pearly shells from the Pierre Shale and its Canadian equivalents. So gorgeous are these fossils that a new mineral industry has sprung up, producing a reddish jewel called *ammolite*. Yet it is not the extraordinary completeness and pristine nature of the fossils that makes them so useful to the geologist (though all of that helps); rather it is their very abundance, and the rapid dance of evolution they seem to have engaged in.

The Western Interior Seaway was a cauldron of evolutionary change. All manner of creatures evolved rapidly in this nutrient-rich inland sea. Ammonites seem have been especially bitten by this itch to evolve, which accelerated the creation of new species. Most speciation is aided and abetted by geographic separation; some small band is split off from the larger population centers and, through time and isolation, gradually (or rapidly) is transformed into a new species, a form no longer capable of successfully reproducing with members of the old group. New species are generally characterized by new morphologies, and so it was with the ammonites, which changed their shells rapidly by adding or subtracting ribs, by altering the spines, and by twisting or unrolling or tightening the loops that defined their shells' shapes.

The great breakthroughs that enabled scientists to use radiometric dating for sedimentary rocks with enclosed ash beds are not limited to Cretaceous rocks. Radiometric dating has now been applied to sedimentary rocks of all ages and has made it possible to assign numerical ages to the entire geological time scale. There are still many uncertainties. But progress has been rapid and the scientific benefits enormous. We can use this information in the search for oil and minerals and to answer questions about rates of evolution. The use of this time machine has been a major milestone in the study of the past.

For work with Cretaceous-aged rocks, the benefits have been enormous. Here, finally, was the realization of D'Orbigny's dream: time, as elucidated by an unbroken sequence of sedimentary rocks sprinkled with fossils and diagnostic ash beds. By interleaving information from the two, geologists put together the highest-resolution scheme of ages found for *any* of the earth's geological time scale units. Because many of the Western Interior ammonites could be found in Texas and even Europe, this time scheme allowed direct correlation with these faraway places. The standard for time, *Cretaceous* time, became centered in the middle of North America. Yet one great problem remained: None of these fossils could be found further west, along the Pacific coast of North America, in Japan, or elsewhere in the vast Pacific Ocean region. To tell the time of these places, including ancient Sucia Island, required an additional step—the invention of an entirely new type of geochronometry device. The age of the Cretaceous on the western margins of the North American continent was decoded by using the earth's magnetic field.

3

Magnetic Clocks

The malevolent California sun stabbed through swarms of insects that late afternoon in the typically hot summer of 1980, its fangs gradually releasing their grip on day. My new graduate student Jim Haggart and I sat in a cool pool, watching amphibious insects bobbing merrily in the rushing creek, and trying not to notice the cheerful skinny dippers several hundred yards down-stream. As we lazed, resting stretched muscles, our ears still rang from a day spent drilling and coring Cretaceous-aged rocks. We were distant from any town, far up a canyon carved deeply into the Sierra Nevada, trespassing like the locals on private land held in trust by a California hunting club—land purchased so that its absentee landlords could slaughter the local deer in privacy. We rationalized our illegal trespass on the principle that science was more important than land rights, but in truth the hunters probably didn't care. They rarely visited their land this time of year and in any case were far less likely to resent our activities than those of the local pot farmers who managed the land farther upstream. All around me I could see the results of our labors: On the edge of the creek were piles of white bags, each containing a tubular rock core about 2 inches long and an inch in diameter. In the bank I could see the scars that yielded up these cores, small holes bored into the solid rock by our gasoline-powered drills modified from chain saws. Also on the creekside lay: cloth sample bags filled with pearly fossils: beautiful coiled ammonites and flat whitened clams wrested by hammer blow from the entombing sediment that makes up the creek bottom and its canyons. It had been 12 hours since we had climbed out of sleeping bags. I suppose I was tired, but how tired could you be at 30 years old, when you believe you are on the verge of putting an accurate age on sediments known to scientists for over a century? How tired can you be sitting naked in a cool creek with nothing left to do that day except scrounge up dinner? We had taken samples that would tie us into a web of research that originated with the classic work done in France and Italy—the regions first studied by the pioneering geologists a century or more before us.

The geomagnetic polarity time scale

The two most common methods for distinguishing units of geological time are using fossils (which yields only a relative age) and measuring the radioactive decay of one isotope into another, which yields an age in years. Yet if you reside in a land where the fossils are unique and new, or where none of the rocks contain the elusive minerals that embody radiometric clocks, you have no way to tell your rocks' age. Luckily for those interested in the age of rocks from the west coast of North America, however, there are other methods as well. One of the most powerful yet least expected timekeepers emanates from the earth's magnetic field. In California, in the early 1980s, I was the first to use paleomagnetism (as this dating system is called) to arrive at an age for the Cretaceous of California, and in so doing I took the first major step toward finding the age of Sucia Island. This tool had been widely used for other time periods and other places, but never before in western North America.

All magnets exhibit the curious property of being bipolar; they all have positive and negative poles. In an experiment beloved of sixth-grade science teachers, iron filings sprinkled around a magnet obediently orient themselves in concentric patterns around these poles. Less obvious, however, is that the poles are producing forces that differ markedly in direction. Why does a compass needle always seek the North Pole rather than simply pointing toward either the north or the south pole? A compass, after all, is nothing more than a small magnet in a closed box, a magnet whose positive pole always seeks the direction of a negative magnetic pole. When placed together, two magnets either cling to or repel each other, depending on whether their positive and negative poles are in contact. Although the forces emanating from these poles are of equal intensity, something about them is markedly different.

The magnets with which everyone is familiar, be they horseshoe-shaped, rods, or the small disks placed on refrigerators, are all quite consistent in their polarity: Their positive and negative poles are always in the same place. But other, more complicated magnets can be made to act in a more variable manner: Their polarities sometimes can be traded, so that the positive pole becomes negative and the negative pole positive.

Magnets come in all sizes. The largest we are acquainted with (though there may be many larger out in space) is the earth. The earth acts as though it had a huge bar magnet in its interior, aligned roughly north–south. At first glance, the earth would seem to be an unlikely magnet: Why does it have magnetic properties at all? Perhaps metal on the earth's surface creates the earth's magnetic field. But closer scrutiny casts doubt on this possibility. Most of the earth's crust is made up of *nonmetallic* material, and thus nonmagnetic material. There clearly isn't enough metal to produce the enormous magnetic field surrounding the earth, a field so strong that it can deflect cosmic rays and solar radiation that would otherwise strike our planet's surface. It turns out that this magnetic field emanates from deep within. Its source is the innermost shell of the earth, a liquid iron core, which behaves like a magnetic dynamo.

Even though the magnetic field of the earth (like all magnets) has two distinct poles, it can be thought of as being created by a uniformly magnetized sphere. The magnetic field pole positions—today—are 79N, 70W and 79S, 110E. Yet these positions are not fixed on the earth. It has long been known that the positions of the magnetic poles change. The magnetic field can be described by three variables: its intensity, or strength at any geographic position; its declination, which is the deviation of the north magnetic pole from the geographic north pole of the earth's rotational axis; and its inclination, or the steepness of the field at any geographic point (If a compass needle is allowed to move freely, it will dip downward, seeking the magnetic pole. The angle of this downward dip is the inclination).

Of these three, declination shows the most movement. The declination of London was 11E in 1580 and was 24W in 1819—a change of more than 36 degrees in 240 years. There are only two ways to account for this change. Either London moved or the magnetic pole wanders across the face of the earth through time. This movement is called secular movement.

The secular movement of the magnetic poles was a key discovery of the nineteenth century. But an even more startling discovery was made in the early twentieth century: The poles can reverse polarity. In the early 1900s, two French geologists, using the most primitive equipment, discovered that the same lava outcrops in France preserved two diametrically opposing directions of polarity. Because the polarity was detected from rocks long since solidified, these earliest paleomagnetic measurements were looking at fossil compasses, where the directions of the magnetic poles could be determined. The fact that two opposing directions were found in the same masses of lava sparked debate that continued for decades. Finally, after repeated measurements with ever-more-sophisticated equipment, there was but one inescapable conclusion: The earth's magnetic field had somehow switched polarity, the positive pole becoming negative, and vice versa. The only alternative explanation was that the whole earth itself had rotated 180 degrees so that the north pole became the south pole. The rotation and orientation of the earth had not changed. The direction of the earth's magnetic field had!

Why reversals occur is a mystery. Many theories abound, and all deal with complex interactions taking place deep within the earth. The earth's

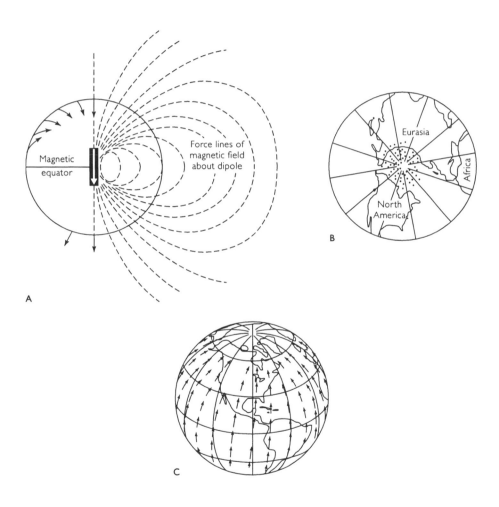

A) Orientation of Earth's magnetic field with the internal dipole coincident with spin axis. B) Magnetic pole position during the last several hundred thousand years. C) position of modern north magnetic pole.

core—the innermost of the three largest shells of earth structure—has never been observed and never can be, Jules Verne fantasies aside. Yet we know a great deal about it, and much of this information comes from the way and the rate at which earthquake waves more through the earth. The core, unlike the overlying mantle, is liquid, but liquid in a hellish way. It is under such high pressure that although we technically classify it as a liquid, it is a type of liquid that cannot exist on the surface. The composition of the core is metallic (it is composed of iron and nickel), and its contact with overlying mantle (which is "solid") must be one of the more interesting places in the solar system. Complex interactions that occur at this core–mantle interface have enormous ramifications for the rest of the planet. The discovery that the core is made of hot liquid metal showed that it is the source of the magnetic field, and the perturbations, eddies, convection currents, or other types of movement within the spinning core may account for the magnetic reversals. Perhaps irregularities or gigantic phase interactions between this liquid core and the overlying, solid mantle region somehow trigger the phase changes. Because these regions are located thousands of miles beneath our feet, the evidence on how and why a polarity reversal takes place is never direct.

Scientists have long known about reversals. It was not until the 1960s, however, that the implications of this discovery for calibrating geological *time* became apparent. It was then that geologists began sampling thick piles of lava flows on the edges of volcanoes. Because each individual lava flow could be accurately dated using potassium/argon techniques, scientists were able to record a relatively precise series of ages for the flows. Each dated flow was then sampled for its *paleomagnetic* direction. To the surprise of the investigators, not only could individual normal and individual reversed directions be detected, but *many* of these reversals in magnetic field were observable. Geologists soon realized that the present, "normal" magnetic field direction has existed only for about the last half-million years. Prior to that, the field was "reversed" relative to its present pole directions. As ever-older piles of lava were sampled, it became clear that the interval between reversal episodes was irregular but generally quite long—on the order of hundreds of thousands to millions of years.

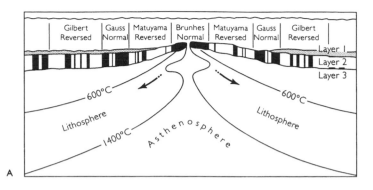

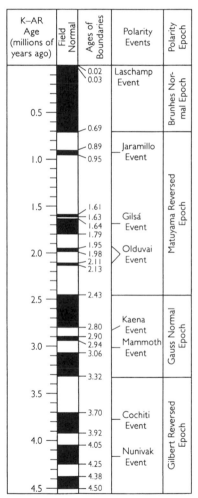

A) Formation of magnetic anomalies in deep sea crust as produced at mid-ocean ridge spreading center. B) Radiometric dates and reversal history for the last 4.5 million years.

At about the same time, oceanographers made a similar discovery for underwater volcanoes. In the early 1960s, plate tectonic theory proposed that new oceanic crust (which is composed of lava) is created by long, linear volcanoes arranged along submarine mountain chains called spreading centers. New ocean crust is then carried away from this spreading center, carried piggyback on a thicker layer of the earth's crust. When oceanographers towed, over the spreading centers, instruments capable of detecting the orientation of the earth's magnetic field, they observed regions of normal and regions of reversed polarity symmetrically arrayed around the spreading centers. The magnetic signals looked like great striped patterns, which were ultimately mapped across all of the earth's ocean bottoms. The only thing investigators needed to obtain a chronology of the reversal history of the earth's magnetic field was an age measurement for each stripe. These measurements were soon obtained by a ship specially designed to sample lava and sediment cores drilled from the bottom of the sea.

By the late 1960s, these cores had provided enough information about age and polarity for a "polarity time scale" to be constructed. The greatest advantage of using magnetic reversals as time indicators is that they are worldwide, or "isochronous," time surfaces. By themselves they are virtually useless; there have been so many reversals during earth history that *no* individual reversal is identifiable. However, combined with other dating techniques, such as biostratigraphy (telling time with fossils) and radiometric dating, the pattern of magnetic field reversals becomes a very powerful tool. Each time a field reversal takes place, it leaves its indelible signature in the earth's history and provides a worldwide time marker of enormous utility. The record of reversal through time is now well known. That accumulated record is called the geomagnetic polarity time Scale, or GPTS.

Detecting the record of these geomagnetic reversals is theoretically simple, but like many *theoretically* simple things, the actual detection process is often less so. The evidence comes from the directions of untold numbers of tiny, magnetized mineral particles locked within either sediment or lava.

The most common of these magnetic minerals, magnetite, is a rod-shaped crystal with a positive and a negative pole, just like any other magnet. Some of these mineral grains are microscopic in size, and if they exist in a medium where they can move freely (such as in water or even in unsolidified lava), they act as tiny compass needles. Their positive pole will point toward the negative pole of the earth's great magnet. If present in sufficient quantity, they endow their enclosing mother rock with a magnetic signal.

Magnetic particles yield useful information about ancient magnetic fields in volcanic and in some sedimentary rocks. Both of these rock types preserve an actual record of the earth's magnetic field direction in much the same way. When hot magma cools and solidifies, the magnetite crystals found within the cooling lava become aligned to the present field direction. A similar thing happens when sediments lithify from a wet slurry to solid rock and, in the process, lock in place the tiny magnetic minerals, all aligned in one direction by the earth's magnetic field at the time of the rock's formation.

These two types of rock now contain weak—but measurable—magnetic signals. *If the exact orientation of the rock is known*, then a piece of the rock, carefully removed and taken to the lab, can yield not only its magnetic intensities but also the actual direction of the earth's magnetic field when the rock was formed. Thus one can learn whether the rock in question was lithified during a period when the north pole of the earth coincided with the positive pole of the earth's magnet, or vice versa.

The pioneering geophysicists, when measuring closely spaced samples from piles of lava, found that there was a random pattern of reversals. Sometimes thick piles of lava recorded numerous reversals; and sometimes an equal thickness—and, perhaps, an equal time period—recorded no reversals at all. The reversal themselves had no identity. Like a computer, they simply record binary data. But just as a simple plus-and-minus computer code can reveal, retain, or record great quantifies of information when enough data are accumulated, so too can our rich record of magnetic reversals tell us much about time.

Synchronizing our watches

An enormous amount of new understanding about the history of the oceans came from study of the deep sea cores collected by the Ocean Drilling Program. We learned that the ocean basins are young compared to continents and that they are ephemeral. Yet although we greatly improved our understanding of the *history* of ocean basins, the deep sea cores did little to improve our understanding of polarity changes, because the soft muck brought up by the drill cores disintegrated easily and so yielded little information about ancient magnetic directions. It was thus virtually impossible to correlate between the reversal records found in this undersea lava samples (with their isotopic ages) and the biostratigraphic time scale made up of land-based fossil occurrences. Correlation between the deep sea samples (with radiometric and magnetic information) and the European stratotype sections (with fossil information) was thus problematic at best.

This problem was especially pronounced for the Cretaceous. Nearly all stages originally defined by d'Orbigny and his followers were identified in shallow-water limestone, which, it turned out, was without any magnetic minerals or even ash beds appropriate for radiometric dating. Even more troubling, the various outcrops were discontinuous; they were isolated in various villages or seacoasts and could not be assembled into any sort of composite sequence or pile. What was needed was a readily accessible record of hard, *magnetically* measurable strata in Europe containing both fossils and a paleomagnetic signal. If a thick sequence of fossiliferous *strata* could be found with a record of the magnetic reversal history, the reversals could be used to determine the age of the strata. Such a record was discovered in the early 1970s, in the Apennine Mountains of Italy.

The discovery of the best region for calibrating reversals and the geological time scale as defined by fossils came about by accident. In his recently published book *T rex and the Crater of Doom*, geologist Walter Alvarez noted that he and co-worker Bill Lowrie undertook the Herculean labor of sampling thousands of oriented cores from thick sedimentary successions of white limestone in central Italy because they were looking for evidence that the Italian peninsula had "rotated" in the geological past during tectonic interactions with Western Europe. Laboratory analyses of these cores, however, showed that there was so much movement of various limestone layers, one upon another, that no detailed history of plate motion could be obtained. What *did* emerge was an unexpected and magnificent record of magnetic reversals. Because these Italian limestones contained a rich record of *microfossils* (mainly the skeletons of single-celled organisms called planktonic foraminiferans), the individual reversals discovered by Lowrie and Alvarez could be correlated both with other land-based sections and with deep sea lava, where reversal boundaries were dated using radiometric techniques. For the first time, European fossils could be associated with radiometric dates by using the magnetic reversal history as a go-between, and for the first time it was confirmed that the detailed and continuous deep sea record of change in magnetic field polarity could also be detected in sedimentary rocks found on land.

The magnetic reversal record discovered in the thick, white limestone of Italy contained several surprises. First and foremost was the discovery that for more than half of the 60-million-year Cretaceous Period, *there were no magnetic reversals at all*. Before this, it had been assumed that the cadence of magnetic pole reversal was rather constant, with a shift every half-million to million years or so, and indeed much of the geological record reveals just this type of pattern. But for reasons still unclear, the mechanism that *creates* magnetic field reversals occasionally went on holiday. From about 118 to about 83 million years ago, the magnetic field was locked into normal polarity. Magnetostratigraphic dating is useless for this interval. However, the first reversal *after* this long interval of normal polarity is one of the most recognizable in the entire geological column. This "long normal," as it came to be called, was followed by a return to reversed and then the restoration of normal polarity in shorter intervals, which produced a readily recognizable "fingerprint" of polarity. Anywhere on earth where you know (from fossils or radiometric dating) that you are in the Cretaceous, and where you have just received a reversed-polarity signal following a long interval of normal polarity, you know precisely where you are in the geological column: at 83 million years ago.

The years of field work necessary to complete the mammoth project of sampling the Italian Apennines yielded more than a detailed look at 100 million years of paleomagnetic reversals. It was from these sections that Walter Alvarez collected the clay samples that yielded unexpectedly high concentrations of iridium—and led to the now-famous hypotheses that the earth was struck by a large asteroid 65 million years ago and that the global environmental effects of that impact caused the Cretaceous/Tertiary mass extinctions, killing off an estimated 60% to 70% of all species then on earth. An animated controversy arose among earth scientists and lasted more than a decade. Was this extinction rapid and catastrophic or was it gradual, lasting millions of years?

These two great pieces of work—the demonstration of a land-based section for studying Late Mesozoic and Cenozoic magnetic reversals linked to biostratigraphy, and the new theory about one of the greatest of all mass extinctions—caused a tremendous surge of intellectual excitement. Its epicenter was Berkeley, where Walter Alvarez worked. The late 1970s and early 1980s were aglow with the excitement generated from these two pillars of research. That glow readily made its way 90 miles east to the University of California at Davis, where I was still assistant professor of geology. I was swept up in both fields, and I remain so to this day.

The extinction controversy was by far the more exciting. How could deciphering the age of west coast rock assemblages compete with dinosaurs being snuffed out by cosmic collisions? Yet the extinction controversy was being played out far away, on outcrops in Europe, and in a sense, the first step to understanding whether the Alvarez theory was correct in determining the age of the rocks in question. Nevertheless, it was with some wistfulness that I resigned myself to working in the wings, while others lept to center stage as the extinction drama unfolded in nearby Berkeley. Chance would soon change that for me.

In the century since William Gabb had made the first discoveries of Cretaceous-aged rocks in California, so much had changed. The Indians were gone, the rivercourses disrupted, the great interior valley turned from rolling grassland to farm. But some things had *not* changed: The California fossils were still endemic, and there was still very little understanding of how to correlate these west coast Cretaceous rocks with the standard reference sections in Europe. Because the fossils in these regions were not of the same species, it was impossible to use fossils as means of comparing the ages of the two regions. Yet the breakthroughs achieved in Europe and by the Deep Sea Drilling Program in the fields of integrated biostratigraphy and magne-tostratigraphy offered a new key to the puzzle. And then one of my new col-leagues at Davis offered the opportunity.

Professor Ken Verosub had graduated in physics from Stanford Univer-sity but had gravitated to the study of paleomagnetics. His initial suggestion that I might incorporate the then relatively new field of magnetic sampling into more traditional biostratigraphic methods got me started. At the time I had absolutely no idea how such sampling might take place. The Alvarez ar-ticles about sampling the Apennine Mountains in Italy mentioned only that "oriented cores" were taken from the rocks. But how? In my experience, cores were usually several inches thick and extracting them required oil-well-drilling techniques. I envisioned large, toothy metal bits being slowly screwed down through rock. Yet, the Italian work had required that thou-sands of "oriented cores" be obtained; and I knew that thousands of oil rigs had not been marched over the Italian countryside. My introduction to the tool designed to obtain oriented cores was thus a pleasant surprise: It was a chain saw, but a chain saw mutated somehow into a coring device. The chain and saw were gone; the motor remained and was used to turn a hollow metal tube coated with diamonds, which cored the rock.

This Rube Goldberg apparatus worked. The long tube attached to the motor was a small diamond drill about an inch in diameter. The motor turns it at high speed, and the thousands of industrial-grade diamonds coating the tip are sufficient to cut several inches into any type of rock. The drill exca-vates a 2- or 3-inch core; this core is then popped off with a chisel. If the exact attitude of the core is known (the direction of its long axis and its angle of dip into the earth are measured with a compass), then it contains all the information necessary for decoding the polarity of the earth's magnetic field at the time the rock was formed. Yet obtaining an oriented core is only the first part of the process of arriving at ancient magnetic directions. Once collected, the core must be analyzed in a large and complex machine called a magnetometer. The intensity of the earth's magnetic field itself is small, and the amount of magnetic signal given off by the magnetite grains found within a core 2 inches long and 1 inch in diameter is minute. The magnetometer was devised to measure these very tiny magnetic fields still existing within the cores.

Like so much in science, the theory is deceptively simple: An oriented core is extracted from some unsuspecting rock and taken back to a laboratory. There it is put into a gargantuan electrified machine out of Baron von Frankenstein's worst nightmare, and *voilà*, a number flashed on some computer screen tells you whether the rocks from which the core was taken were deposited when the earth's polarity was normal or reversed. And, as in a low-budget ad on TV, you get the something else thrown in absolutely free—in this case, a measurement of the field intensity from the rocks you have sampled and the inclination and declination of the magnetic field from the age and locality you have sampled. These latter bonuses can provide some of the most interesting information of all; they can tell you the ancient latitude of the rocks you have sampled. I was seduced. At the time, no one had sampled Cretaceous-aged rocks for western North America, and the whole procedure seemed so straightforward: Drill some cores, analyze them, and end up with a polarity history that could be matched to the Italian work completed by Lowrie and Alvarez. Why not? What could go wrong? What could be easier than deciding in a binary system? All one needed to find out was whether the fossil magnetic signal in the rocks reported a normal or a reversed polarity for any given time.

Who has not been so seduced, be it for a used car, a VCR, or a new partner with no baggage? I took the bait, swallowed it, and headed off into the California hills with all the equipment needed to drill hundreds of small cores from the rocks. And everything worked—the first few times. (That's the secret of a really effective con: Make it work—at first.)

It takes two to drill—one to do the actual drilling, the other to pump the water. Water? Somehow the water hadn't come up when I received my first sales pitch from Dr. Verosub, who, like any good used-car salesman, also declined to spend any time in the field actually drilling the rocks after our first trial run. The drilling requires that copious quantities of fresh water be pumped constantly through the drill while it is in contact with the rock. Spinning at propeller-like speed, it also emits the screeching characteristic of chain saws (but elevated in pitch because of the diamond-on-rock contact). I quickly found that coring rocks for paleomagnetic sampling was anything but a rest cure. The drill had to be muscled into place, and the coring of the rock, accompanied by the constant stream of water being pumped into the hole, enveloped the operation in a fine stream of muddy haze. The driller was soon coated with wet mud. Three cores were taken at each locality, and we drilled a new locality in each meter or in several meters of layered rock. Because our first sampling creek, Chico Creek, was over 500 meters thick, we were soon locked into a long-term endeavor. On a good day we could obtain about 100 cores, but we often got less.

Yet the drilling turned out to be the *easiest* part of the operation, for once the holes were bored into the rock, the cores had to be extracted and their orientation noted. Orientation was found by inserting a brass sleeve around the core (which, if we were fortunate, was still in the rock, attached at its base, and not broken off and jammed inside the drill). The sleeve had a platform on top, which was machined so that it held a geologist's compass. When leveled, the compass showed the orientation of the core, and a second measurement yielded the plunge, or the angle of its entrance into the rock. Given these two observations, coupled with a measurement of the orientation of the sedimentary beds themselves (which sometimes were in their original, flat-lying orientation but more often were tilted at some angle, a complication that had to be accounted for), a computer could calculate the original orientation of the core.

All this took time—time to select a site, time to drill it, and much time to take the accurate measurements necessary to arrive at core orientation.

Because the drill was gas-powered, it needed constant refueling and lubrication. The drill bits tended to break. The water can was always in need of refilling. Even so, the hard work was relieved by its taking place on one of the most beautiful stretches of land in the world.

Chico Creek had been spared most of the ravages of the Gold Rush. That event, even though it occurred nearly 150 years ago, left ugly blemishes on the rivers of California that still persist. In their mad quest for gold, the miners dredged tons of gravel and rock from the state's rivers and left these waste piles everywhere in the river valleys. The Mother Load, in the foothills east of Sacramento, was hardest hit, and those rivers of the north, such as the Feather and Touolome Rivers, were devastated as well. Chico Creek is a smallish creek flowing out of the Sierra Nevada foothills. It too had its share of gold, which, because of the rugged country around, was mined by hand rather than by machine dredges. But it has other treasures as well, treasures first discovered by William Gabb. It has Cretaceous-aged fossils of spectacular beauty.

The geological setting of Chico Creek is itself extraordinary. Most of the Sierran foothills in northern California are covered with drab rock: thick, gray volcanic residue from the Sierras uplift blanket the older rocks in the region. Only where creeks and rivers have cut deep gorges can the more ancient geology of this region be found. On Chico Creek, the high valley walls are made up of the thick volcanic rock, and only in the creek itself and in the low walls lining it are different rocks found, sedimentary rocks of much greater antiquity. Here the creek has knifed down into an ancient sea bottom, where clams and snails and ammonites in untold numbers lie entombed. The cool creek burbles merrily over this smooth, dark rock with its myriad white fossils, and each year the flooding snow melt erodes the surrounding rock to bring new treasures into view, treasures that sparkle briefly before they too are eroded into sand grains on their way to the Pacific Ocean.

Gabb's pioneering expeditions to Chico Creek in the earliest years of the California Geological Survey showed the presence of Cretaceous-aged fossils in California, as we noted Chapter 1. Although Gabb could find no fossils similar to any species known from Europe, the prints of numerous ammonites belonging to the same genera, if not the same species, as European forms convinced him that he was sampling not just Cretaceous-aged rocks, but *upper* Cretaceous rocks, equivalent in age to the chalk of the English and French coastlines. His collections were hurriedly made; he had, after all, an entire state to investigate, and no one portion, however physically beautiful, could monopolize his time. He hoped that subsequent expeditions would reveal some ammonite species occurring in both France and California, which would allow those two distant relics to be tied together by a filament of time. It was not to be. In the decades following Gabb's pioneering studies, the countless scientists and amateur fossil collectors who had been drawn to the deep valley of Chico Creek by its rich fossil content searched in vain. For all the great collections subsequently amassed, no investigator stumbled on any fossil diagnostic of the European Cretaceous. The great rampart of North and South America apparently created an impassable barrier for European species: They could not colonize the new world of the Mesozoic Era. No species in Europe exchanged DNA during the Mesozoic Era with any species in western North America, just as no mollusk species in these two areas do so today. If you take a paleontological specialist to see fossil collections made from deposits in the Pacific region, he or she will look in vain for old friends and will confront only strangers. Yet while the ammonites and other diagnostic European fossils migrated westward with the currents over their myriad generations (ultimately to find more westward expansion blocked by geography), the earth's magnetic field created an indelible time marker in the Chico sedimentary rocks.

Through a great portion of the Cretaceous Period, which has been radiometrically dated as beginning about 145 million years ago and ending 65 million years ago, the earth's magnetic field was stuck in one direction. There were no reversals for tens of millions of years. And then, about 80 million years ago, this long quiescent period came to an end with a reversal. Although any reversal is a worldwide event, there have been so many through earth history that a reversal is essentially useless without other, time-bound data such as those obtained through fossils or radiometric decay. But the first reversal of the Cretaceous "long normal" episode was a diagnostic fingerprint that could be used with only a minimum of other age control. All you needed to know was that you were sampling somewhere in the upper Cretaceous Period—a time interval itself more than 30 million years long. If you then encountered a zone of reversed magnetic polarity sitting atop a long interval of normal polarity, you could be assured of having found the single most age-diagnostic magnetic event in the history of the earth. This magnetic reversal was given the unromantic name of 33R. It was known from the deep sea and from the Italian Apennines Mountain work. In 1980, analysis of my drill cores from Chico Creek showed it there too. Like an elusive fish finally hooked, the Cretaceous of the west coast of North America now had a line—in this case a time line—securely attached to the global time scale.

According to Ken Verosub, who analyzed the cores drilled from Chico Creek sediment in his paleomagnetics lab at the University of California at Davis, the magnetic signal captured in our cores was strong. But was it accurate? Did the signal emanating from these rocks record the directions of the earth's magnetic field from a time when dinosaurs roamed the earth in profusion, or was it from a more recent time? In the lab, I learned that paleomagnetics is fraught with uncertainty and that the black and white reversal patterns shown on all summary charts of the earth's magnetic field history—apparently so clean and unambiguous in their binary glory—are really anything but.

Like so much else, paleomagnetics sounds simple in theory. The sediments we had drilled from Chico Creek were sandstone and siltstone deposited on shallow sea bottoms long (but exactly how long ago?) ago. Among the minute sand and clay particles drifting down onto that ancient sea bottom were trillions and more of magnetite grains; all were eventually squished together into the bottom sediment. The last thing these tiny magnetite grains did before being cemented into place was orient themselves parallel to the earth's magnetic field. Each core so laboriously drilled from the rocks held some number of magnetite grains, and each core thus exhibited a tiny but measurable magnetic field of its own, created by its enclosed magnetite. Fed the orientation of a core, the magnetometer could tell us the orientation of the magnetic field emanating from it. One by one, the inch-long cores were placed in the magnetometer, itself a curious contraption of iron and wires resembling a 6-foot rounded coffin. Eventually three numbers emerged: the strength of the magnetic signal carried within each core and (when the various attitudes of the core and the sedimentary beds were sorted out by the computer running the magnetometer) the declination and inclination of the earth's magnetic field at the time the rock that had been cored solidified. These precious numbers revealed whether the field at that time was normal (like today's) or reversed. The numbers from Chico Creek showed both normal *and* reversed polarity. I was jubilant: We had made the first discovery of polarity reversal in Mesozoic-aged rocks collected from the west coast of North America. But were these results "real"? Were they indeed a magnetic signature frozen in stone since the Mesozoic Era, or were they merely an artifact of later magnetism imprinting itself on these rocks?

This was the first inkling I had had that paleomagnetism was not so simple a tool to operate as I had anticipated. It was explained to me that the magnetic signal frozen in the sediments we had sampled—or in any other rocks, for that matter—maintained its original direction only as long as the rocks were neither reheated nor subjected to a great deal of ground water passing through their pores. If the rock had been reheated enough, the tiny magnetic rods of magnetite that had so faithfully recorded the ancient magnetic field present at their consolidation into this particular sedimentary rock would have reoriented themselves to the direction of earth's magnetic field at the time of reheating, and they would have done so without leaving any clue. There would be no sign saying, in effect, "We have been reheated! We now give spurious results! Paleomagnetist beware!" Huge mistakes can be made. They have been made and will be made again.

And is that the end of it? No, there is more. Much of the time, the rocks you sample have been heated, all right, but not quite to that magic temperature (called the Curie point) where the magnetic direction resets. Instead they take on a slight, but not complete, overprint of new magnetic field direction. In these cases the investigator must strip off this overprint, either by progressively heating the cores in an oven and running them again, or by subjecting them to an alternating electric field that (sometimes) accomplishes the same result. You might say the paleomagnetist has to cook the data. And there is still more! Lots of rocks have no magnetite, or magnetite is present in such vanishingly small quantity that the standard magnetometer cannot detect the ancient magnetic field. In this case a new instrument must be employed, a superconducting instrument that has extremely high sensitivity.

Thankfully, the Chico Creek cores showed no such problems. According to my learned paleomagnetist colleagues, they "behaved" beautifully: They had a strong signal, showed both normal and reversed polarity, and seemed to have no overprint. With minimal laboratory effort we attained publishable and very interesting results. We were able to identify the Cretaceous "long normal" and the first reversal above it. For the first time, rocks from the west coast of North America could be reliably dated and correlated with European stages and deep sea sediments. I was converted. This method seemed too good to be true. You guessed it. It was.

Over the next two years, Jim Haggart and I scoured the eastern and western sides of the Great Valley sequence, drilling rocks and collecting ammonites. We helped tie the local California strata and its enclosed ammonite biostratigraphy to the international time scale. There were no ashes for us to date, so there was no way to find absolute ages. But the long normal polarity followed by this first reversal was a time marker almost without peer in the geological record. Soon after our confirmation of this reversal in Cretaceous rocks of northern California, my colleague David Bottjer and others from southern California located it as well. I was spoiled by this early success.

We seemed to be on the verge of a major breakthrough in telling time in western North America. But the process was so convoluted! We needed to compare the age of the rocks in California with those of the Vancouver Island region and then compare both with the regions in Europe where the various time units were first identified. These European "standards" of comparison are called stratotypes. We could do this if we could find ammonite or other fossil species that had lived in both places. But none had! We ourselves could make the comparison if we could find rocks in both regions that could be radiometrically dated with a mass spectroscope. But neither Europe nor the Vancouver Island region had the necessary bentonites (ash layers) interbedded with the sedimentary rocks, so that approach was out. Our last hope was the use of paleomagnetics. But once again we were stymied. The rocks in Italy that Walter Alvarez and his colleagues had drilled had no ammonites. To complete the circle, to make possible the correlation, someone had to complete paleomagnetic analysis on European rocks with both microfossils and ammonites and then complete the same type of paleomagnetic sampling in the Vancouver Island region. Only one suitable place was known to me in all of Europe: a region I had first visited in 1982 while researching another scientific question, the reason for the mass extinction at the end of the Cretaceous Period. This place was a seacoast section in Spain known as Zumaya.

Thus an ambitious plan was born. I decided to head east, to Zumaya, to core rocks in the classic European regions and then head northward and try it out on the rocks of the Vancouver Island region, including Sucia Island. Little did I know that I was approaching a problem already ruled intractable by workers far more experienced in the way we do magnetic rock analysis. I had enjoyed beginner's luck with the paleomagnetic time machine. I was about to learn that this particular time machine was a very delicate apparatus indeed.

Calibrating our clocks

We struggled into the Madrid airport in early May of 1984. Having taken an all-day flight from San Francisco to New York and then having flown all night from New York to Madrid, my friend Jeff Mount and I, a two-man expedition, numbly dragged a heavy wooden crate from the international to the domestic air terminal. The crate held two paleomag drilling units and all the ancillary and peripheral gear needed to conduct that research. Our goal on this trip was simple: Retrieve a detailed history of the magnetic reversal stratigraphy from a classic outcrop at Zumaya, Spain. Here we would be duplicating the paleomagnetic records of this region made in Italy by Walter Alvarez and William Lowrie a decade earlier. But our rocks contained something precious that is wanting in Italy: larger fossils, including ammonites.

Here, if we could obtain a reliable paleomagnetic record, we could correlate the biostratigraphic markers of the latest Cretaceous, as recorded by ammonite fossils, with the paleomagnetic reversal record.

We had boarded the plane on a breathtakingly hot California day, and Jeff traveled in shorts. By New York he was shivering, by Spain a vicious cold had taken him—a rhinovirus assault of some gravity. But who in his thirties surrenders to a cold after traveling thousands of miles to finish a great deal of work in a short time? As the days went by he got sicker and sicker, and still we worked. Up at 6 each morning, the Spanish excuse for a breakfast (strong coffee and a length of French bread and butter), and then onto the outcrop. Through the coastal village of Zumaya, we guided our small rental car through the tiny lanes scraping by old brick and plaster buildings, past the haystacks to the isolated coastal exposures. Drag the drill onto the rocks, fill the gas tank and the large water canister necessary to cool the drill, find the last site from the day before, and begin drilling anew.

Spring is the time of squalls in the Bay of Biscay, rapidly moving blusters of wind and rain gathering from the south and belting the coast. We would try to find shelter at first, but eventually we just kept working, huddling underneath our raincoats during the worst of squalls, hoping for the brief islands of sun in the ocean of clouds above. Day by day our sites marched down the beach—and down section. We started at the Cretaceous/Tertiary boundary, the site marking the great mass extinction that ended the Age of Dinosaurs, a site visible at Zumaya as a distinct lithological change. And then we dropped, layer by layer, into ever older rocks, drilling every few meters going down this enormous column of stacked limestone and marl, passing down a million years, and then two, heading layer by layer, page by page of this stratal book, toward ever older oceans, eventually reaching rocks I thought to be the same age as Sucia Island, but who knew? With these cores, however, *we* might know. We would be the first; we would tie Gubbio to Zumaya to Sucia Island in a web of magnetic correlation and time lines.

For 10 days we drilled the coastal exposures. There was other work as well, of course. We collected fossils and we studied the Cretaceous/Tertiary boundary, searching for evidence that a large cometary collision 65 million years ago did indeed end the Age of Dinosaurs. But the primary goal here was to bring home several hundred magnetic cores for paleomagnetic analysis.

We flew home, finally, laden with rocks and the many cores. With impatience I awaited the paleomagnetic results. First, the cores had to be trimmed, and then one by one they were fed into the magnetometer, over and over, as many as 10 times per core. It soon became discouragingly clear the these cores were far more compromised by magnetic overprint than anything we had collected in California; in fact, it turned out that *none* of the cores was useful for paleomagnetics. These rocks had apparently been reheated, the small magnetic crystals in the cores realigned with a much younger magnetic field. The weeks of work were for naught. We had hoped that the Spanish rocks would have the same pristine magnetic signal as those from California, but we were wrong. Several years later, Dr. Jan Smit of Holland would try again to drill Zumaya. I told him about our disappointment. Yet he too refused to believe that these rocks would not yield good paleomagnetic results. He too understood the importance of learning the position of the paleomag reversal in the uppermost Cretaceous at Zumaya. And he too would conclude, as I had, that a magnetic overprint had forever destroyed the primary magnetic signal here. There is no publication of negative results of this kind.

The work in Spain taught me a hard lesson. Magnetic clocks are reliable only in special circumstances. When they work, they are superb. But there is no way you can go up to an outcrop and be sure that you will obtain results.

By the end of 1984, a magnetic reversal had been identified from the Cretaceous of California. But the Nanaimo Group still tightly guarded the secret of its age. Other workers before me had tried to extract its magnetic time secrets; others had tried to use the time machine of a magnetic clock. No one had succeeded. Spain had taught me caution, and I wondered whether magnetic sampling of Vancouver Island was worth the effort. But the reward if it worked! Magnetic analyses tell you more than time: They can pinpoint the ancient position of the rocks you study, the latitude at which the sedimentary or volcanic rocks solidified. They can be the best tool of all for fixing *place*. The rocks we had drilled in California and Spain had

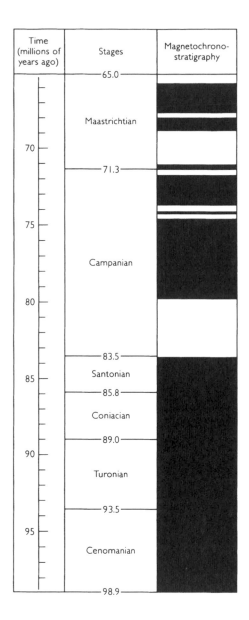

Correlation of Cretaceous stages, reversal time scale, and radiometric ages.

not moved since they were deposited long ago in the Cretaceous, many previous investigators had concluded. Such was not the case for the Vancouver
Island region. During the 1980s it was proposed that the Nanaimo Group
may have originated far to the south of its present position and moved up the
coast by plate tectonics, the mechanism that causes large blocks of the
earth's crust to travel vast geographic distances over time. The best way to
accept or refute this—the best time machine to use—was again paleomagnetics. Now I had two reasons to succeed in my drilling plans for the Vancouver Island region. I still wanted to know about time. But now I wanted to
know about *place* as well.

The story of the paleomagnetic analysis of the California Cretaceous-
aged strata is but one example of how this routinely used method unfolds.
Most such studies proceed in fits and starts, just as ours did. Only slowly
does information accumulate, but it yields some of the best ways in our ar-
senal of time machines to find the age of rocks. Magnetostratigraphy, as this
method is called, is now used on virtually every time interval of the geologic
record.

The solution

Twenty years after beginning a seemingly simple quest, I could finally propose an age for the Sucia Island fossils. The Sucia strata bear the ammonites
Baculites and, more important, a heavily ornamented ammonite called
Hoplitoplacenticeras. These tell us that Sucia Island is Campanian in age and
therefore is of the same age as the rocks in western France where the world's
finest Champagne is created. The species of *Baculites* is *B. inornatus*, which
is also found in California. In a few creeks, this ammonite is found at the top
of magnetozone 33R, the first reversal after the Cretaceous long normal. In
the Western Interior Seaway, this same reversal has been discovered in what
is known as the Middle Campanian. Its upper parts are filled with bentonites
that have been dated between 81 and 79 million years in age. This, then, is
the age of Sucia Island.

Part Two

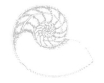

Place

4

Baja British Columbia

The surface of the earth seems so fixed. And it is, in our short lives. But if we take a much longer view, we see that the earth's surface is actually quite mobile. For any locality on earth we can ask, "How long has this place been at this place?" For some decades now the Vancouver Island region, this book's long-running example of how the various time machines work, has been suspected of being a "suspect terrane"—a place that arrived at its current geography through continental drift. And as it turns out, that suspicion was well founded. The story of how we reached this conclusion—the time machine we used to do so—is the subject of this chapter. The story, from my point of view at least, begins on a small Canadian island north and west of Vancouver.

Hornby Island

Islands always have their own flavor. Much of it comes from such disparate elements as topography, climate, vegetation, and fauna. But the human inhabitants also subtly (or not so subtly) change the feel of an island, especially if it becomes a haven for the iconoclasts who sometimes make this world so interesting. Hornby Island of the Canadian Gulf Islands is definitely such a place. It is situated between Vancouver Island and the mainland of British Columbia, a medium-sized isle about 5 miles across. Like so many other of the islands in this region, it lies in rain shadow and hence enjoys some of the best climate (and least rainfall) of any region of British Columbia. And perhaps for that reason alone it attracted a great throng of the artists, hippies, nonconformists, radicals, and dropouts who have made this and the other islands in the lee of Vancouver Island the California of Canada.

My own first visit to the place was in 1974. Along the north shore of Hornby Island, a thick assemblage of nearly flat-lying mudstones contains some of the best-preserved ammonite fossils in the world, and I longed to see these famous beds, with their famous fossils. I was not disappointed. Never really common, the ammonites from Hornby Island occur in nodules, or concretions, and the way to find them is to crack open these nodules. Unfortunately, only one nodule out of many holds an ammonite (or some other fossil).

They are young for ammonites. The quiet muddy sea bottom from whence they come was deposited some time after the Sucia Islands beds and perhaps—perhaps—about 5 million years prior to the great comet impact that wrecked the Mesozoic world, killing dinosaurs, ammonites, and much else in the process. Yet that event was still far in the future when the ammonites lived and died in the Hornby region. The Hornby Island ocean was an ammonite heyday, a last hurrah. On land the great dinosaurs still held sway, and the first *Tyrannosaurus rex* and *Triceratops* were evolving as the Western Interior Seaway of North America gradually receded from the continent. In the contemporaneous late Mesozoic oceans, shelled cephalopods and mosasaurs, archaic fish and bizarre flat clams, tropical reefs made of clams instead of corals—creatures great and small were *alive*. Now they are but stony memories.

My first trip to this island was brief: I blasted the concretions with my hammer, I found a few beautiful fossils, I spent a glorious day on a glorious beach, and I left, just a solitary day-tripper. I would not return for 12 years. I had gathered some fossils, but I had not done any science. I was not really interested in the age of Hornby Island, or in much else about it, at the time of my first visit. That state of affairs was to change greatly by my next.

By early 1986 I had already learned that paleomagnetism, seemingly such a simple and straightforward endeavor, was fraught with difficulty. If the sampled rocks had been reheated to any significant degree, you could not expect reliable results. Those scientists who had sampled for paleomagnetism in the Cretaceous-aged strata of Vancouver Island concluded that the entire region had been reheated and was therefore unsuitable for paleomagnetic work.

But had the previous workers exhausted all avenues? Had they shown without a doubt that no region of the vast Vancouver Island area was unaffected by regional heating? I thought that the easternmost islands of the Georgian Strait might possibly be unaffected, because the ammonites there were absolutely pristine. Like Sucia Island, Hornby had preserved its fossils well. Might not it have preserved a primary magnetic signal too? It was worth a try, so I journeyed there with two student field assistants to drill for paleomagnetism.

The trip took place in perfect weather, a late summer lark. We arrived on Fossil Beach, repository of so many exquisite ammonites, and began the laborious process of drilling and recording our results. Such activity certainly drew attention. Screaming like a banshee, throwing streams of muddy water out of the nozzle, a paleomag drill set is a raw wound walking. Lots of natives came by to see what in the world three mud-spattered men were doing on their beach.

Island folk being the friendly people they usually are, we were soon befriended by a long-legged woman, who graciously invited us to camp on her nearby property. Early each morning we ventured from this cozy spot, packing lunches and dirty clothes, and managed to remove more than a hundred oriented cores, each about an inch long, from the solid rock that makes up the fossiliferous shale of Hornby Island. On the third day the infernal din of our drilling was overshadowed by an even louder noise: A Royal Canadian Mounted Police helicopter landed on our camping spot and proceeded to root out much of the shrubbery. It was the annual Hornby Island pot raid.

Months later the slow analysis of the magnetic cores was completed, and it was as if we had been smoking the pot growing so plentifully on Hornby Island, for many of the magnetic numbers seemed to make no sense. Cores taken from adjacent beds gave wildly fluctuating results. Some cores were somewhat "well-behaved" (the euphemism employed by paleomagnetists for magnetic results that show some semblance of orderly change when analyzed). Most, however, were far more chaotic. According to Ken Verosub, who had been involved in my earlier work in California and Spain, the numbers were too scrambled to be publishable. I thought that more than a few of the cores betrayed the presence of the telltale reversals, but I was overruled. Even more vexing, the grant that funded this project expired, and no more analysis could be run. Once again I had completed a project in which all of the immense work involved in collecting the cores was for naught, or so it seemed. Our results from this expedition were never published. I had to leave the Vancouver Island rocks where they lay. But the revolution in the earth sciences called plate tectonics, or continental drift, was about to make Hornby Island a pivotal place in one of the most interesting of all scientific controversies. Central to this would be the magnetic record from Hornby—and yet another time machine.

Coincidence or continental drift?

The green, fossiliferous shale exposed on Sucia, Hornby, and the other islands of the Vancouver Island region—rocks grouped together by their common age and origin—make up a terrain that in many ways seems iconic of the Pacific Northwest. Who could deny the nativeness of this land drenched with rain, carpeted with fir trees, ferns and clinging moss, and surrounded by cold green seas once filled with salmon and halibut (riches now supplanted by fishing boats chasing a dwindling stock)? Yet far to the south, along the sun-drenched coastline of Baja California, lie sedimentary strata identical in lithology and fossil content to the drab green mudstones so typical of the Upper Cretaceous of the Vancouver Island region. Coincidence? Or continental drift?

Continents do drift. Who would deny the ancient love affair of Africa and South America, hoary continents that long ago dallied in their global bed until they were torn asunder by a restive earth. Cerebral researchers needed complex machines to confirm the theory of continental drift. But the ultimate evidence stares out at us from any map. We know now that the continents lie embedded in a pavement of more dense oceanic crust and that both crust and continents skate across the surface of the earth on a more astringent and liquid interior. These truths are now accepted by all earth scientists (the last holdout, Al Meyerhoff, has died). This revelation ushered in one of the most profound scientific revolutions of our time, rivaling those sparked by DNA in biology and by relativity and quantum mechanics in physics. And as often happens after scientific revolutions, once the battle was over and the defenders of continental immobility routed through evidence marshaled from paleomagnetics, paleontology, and paleoceanography— in short, once the proponents of continental drift had won—the victors, like so many conquering armies, claimed the spoils of war and went a little overboard. In the aftermath of victory, *every* feature on the earth was seen as resulting from some plate tectonic process. If continents could move, why not every other feature on this wrinkled and ancient earth? A blizzard of speculative papers about ancient continental configurations blanketed the landscape. Hypotheses outstripped the ability, and even the desire, to test them. Backlash set in.

Quite properly, earth scientists began to demand convincing proof rather than merely hypotheses concerning continental drift and the ancient continental positions. Yet proof was often quite difficult to assemble. The original evidence confirming the theory of continental drift came mainly from ocean basins, not continents; it was the ocean basins that revealed their long, linear underwater mountain chains, the spreading centers where new oceanic crust is formed. These same ocean basins that yielded the magnetic lineations on the bottom of the sea, showing that ocean crust is ever more ancient as we move away from the spreading centers. It was the ocean basins that revealed the presence of subduction zones, the linear regions where the conveyor belt of crustal movement dives under continents, to be consumed by the earth's interior heat and then move back toward the spreading centers along enormous lithic convection cells that power continental drift. From the continents far less information is available. At least as far as proofs of continental drift are concerned, the best evidence comes mainly from the time machines of stratigraphic analysis and paleomagnetics.

Stratigraphy is the simplest branch of the earth sciences, at least in principle. Stratigraphers study the sequence of rocks as they pile one upon another through time. Sometimes a stratigraphic succession is created by the simplest action of one sand grain falling upon another over time, leaving behind sedimentary beds, oldest on the bottom, youngest on top. Sometimes things are far more complicated, and great slices of earth's crust are thrust over the tops of other huge slices. Yet even in these complex tectonic cases, the order of things is deciphered via stratigraphy.

One of the best ways to detect whether one hunk of rock on a continent is related to another—or even to rocks on a different continent—derives from stratigraphy. For example, one of the first (and still among the most powerful) lines of evidence used by early believers in plate tectonics came from stratigraphic studies conducted in the Southern hemisphere. Investigators such as Alfred Wegener and Alexander du Toit recognized that similar successions of sedimentary rocks were observable on the now widely separated expanses of Australia, Africa, Asia, India, South America, and Antarctica. Long before any believable *mechanisms* had been proposed to account for continental drift, Southern hemisphere geologists defined a diagnostic, quarter-billion-year-old succession of strata (which they named Gondwana succession) and argued (mostly to deaf ears) that such similar stratigraphic successions on continents now-separated could not be coincidence. All of the continents bearing these rocks *must* have been part of a single giant continental land mass at the end of the Permian Period, about 250 million years ago. And so they were.

The concept of continental drift was first given serious consideration in the late nineteenth century, when the then-famous Austrian geologist Eduard Suess suggested that Africa, Madagascar, and India were once all joined together as a single land mass and only later drifted apart. Suess based this heretical proposal on the great similarity in the rock types found in all three areas. He named this ancient continent Gondwanaland, borrowing the name from an area in India inhabited by a tribe called the Gonds. Suess was no charlatan or crackpot, and soon a few other geologists, mainly those working in the Southern hemisphere, began considering the possibility that a "supercontinent" existed there during the late Paleozoic and early Mesozoic Eras.

The various threads of evidence supporting the concept of an ancient, southern supercontinent were woven together in a remarkable book published by the German meteorologist Alfred Wegener in 1912. Wegener was convinced that the similarity in coastlines between western Africa and eastern South America went far beyond coincidence. He amassed paleontological and geological information to support his cause, but his book was met with instant criticism. No mechanism for such "continental drift" could be postulated.

In the early 1900s a young South African geologist named Alexander du Toit began to crisscross South Africa. He was to spend 20 years examining rock structure, mapping huge expanses of territory, and, in the process, filing vast amounts of information away in his encyclopedic memory. Du Toit soon realized that Wegener's outrageous hypothesis explained many of the geological features of southern Africa, and in 1921 he published his first paper about the possibility of "continental sliding." Geologists had long been puzzled about the origin of the mountains rimming the South African coastline. Du Toit realized that they could have been compressed by continental collision; he had a vision of southern Africa caught in a monstrous vice between South America and Antarctica. Du Toit was able to visit other southern continents, where he observed remarkably similar successions of rocks. It was not just that the rock *types* were the same; more convincingly, they showed a similar stratigraphy.

Du Toit went far beyond Wegener in his understanding of Gondwana-land; he was able to imagine both the early merging of the various continental pieces and their climactic melding into a single continental mass in late Paleozoic time, followed by their fragmentation during the Mesozoic and Cenozoic Eras. He became well versed in the stratigraphy not only of his own continent but of the other members of Gondwana as well. Perhaps his most telling argument supporting continental drift was his demonstration of the remarkable similarity among late Paleozoic rock sequences on the various continental pieces. On each he saw a basal unit of rocks deposited during Ice Ages, known as glacial tillites, overlain by lake deposits containing a small aquatic reptile named *Mesosaurus* and then by deltaic and river deposits; the units culminated in Mesozoic-aged basalt. Du Toit called this the Gondwana System, known today as the Gondwana Sequence.

Wegener and du Toit turned out to be correct. But like Vincent van Gogh, who never saw his genius acknowledged, neither Wegener nor du Toit lived to see their great triumph of observation and reasoning confirmed. The proof that continents wander did not burst into the scientific consciousness until the early 1960s, when a slew of studies brought the theory of a static earth tumbling down.

First, studies on rock magnetism showed that either the geomagnetic pole had moved through time or the continents had. Both seemed utterly impossible. But in short order, new evidence supporting continental drift came to light. It was demonstrated that the mid-Atlantic ridge, a then poorly known line of undersea mountains running down the middle of the Atlantic Ocean, was composed of a chain of *active* volcanoes constantly in the process of creating new oceanic floor. Next, a newly instituted program of deep-sea drilling demonstrated that the age of the ocean floor *increases* as one moves away from these submarine volcanic chains, which oceanographers began to call "spreading centers"; this discovery proved that the sea floor *was* spreading and that, in many cases, it carried continents along for the ride. But where was all this new ocean floor going? Seismic studies then showed that in many places on earth, oceanic crust dips downward into the earth itself along long linear arcs of the earth's crust called subduction zones. Subduction invariably leads to mountain chains and active volcanic mountains inland of these features. A scientific revolution had occurred within a few short years. We now know that continents indeed have drifted and that they drift still. They do so because they float.

All continents are masses of relatively low-density rock embedded in a ground of more dense material; continents essentially float on a thin (compared with the diameter of the earth) bed of basalt. Earth scientists like to use the analogy of an onion. The ocean crust can be likened to the thin, dry, and brittle onion skin sitting atop a concentric globe of higher-density, wetter material that itself is moving relative to the much thicker interior of our global "onion." Continents are like thin smudges of slightly different material embedded in the onion skin. Unlike an onion, however, the earth has a radioactive core and constantly generates great quantities of heat as the radioactive minerals, entombed deep in the earth, break down into their various isotopic by-products, liberating heat in the process. As this heat rises toward the surface, it creates gigantic convection cells of hot, liquid rock in the mantle—a molten layer of material directly beneath the outermost region of the earth, the crust. Like boiling water, the viscous upper mantle rises, moves parallel to the surface of the earth for great distances at rates of several inches each year (all the while losing heat) and then, much cooled, settles back down into the depths of the earth. These gigantic convection cells carry the thin, brittle outer layer of the earth—known as plates—along with them. Sometimes this outermost layer of crust is composed only of ocean bed. Sometimes, however, one or more continents or smaller land masses are trapped in the moving outer skin. This process, which is termed continental drift or plate tectonics, is one of the great unifying theories of all science.

The spreading centers produce new crust as giant plates drift apart. Yet these plates are not always diverging. They also converge, sometimes causing the collision of great continents. The collision of two continents is a slow, majestic process each land mass moves at only a few centimeters per year; so thousands of lifetimes must pass before any change in position would be apparent. But as millennia pass, the continents do move relative to one another, and sometimes they collide. The first contact of the opposing continental shelves does little. But year after year, as the two giant continental blocks of earth's crust coalesce, enormous forces of compression act on the continental edges until the outermost regions buckle. Mountains begin to form along the two edges as the collision progresses, often creating high, spewing volcanoes amid the contorted mass of sediment and rock that was once a tranquil coastline of wide sandy beaches. Finally, the two continents are incapable of further compression, yet they are still driving against each other. Slowly, one of the continents slides over the other, often doubling the thickness of their crustal edges in the process; then they lock together, no longer able to give any more ground. A relatively recent and dramatic example is the collision of India and Asia. Forty million years ago, India was a small fugitive of the ancient Gondwana supercontinent, fleeing northward from its origins in the Southern hemisphere until it collided with mainland Asia. In the process, the edge of the Indian continent rode up onto the Asian mainland, resulting in the formation of the world's highest mountains, the Himalayas, which are composed of the thickest continental crust known on earth.

Geologists know now that plates, those enormous lithic fragments that do the drifting of continental drift, can interact with other plates in only three ways: They can pull apart, smash together, or run side by side. The first two phenomena—the divergence of plates at the mid-ocean ridge spreading centers, and their convergence at subduction zones—were the subject of the first pioneering wave of plate tectonics research, which demonstrated the reality of continental drift. Yet the third, the side-by-side motion of plate margins, is equally important. The San Andreas fault in California is perhaps the best example of side-by-side motion. The San Andreas has been operating now for millions of years, and as the thin slice of California on the outboard side scrapes northward, earthquake by earthquake, it separates rock units that once were geographically together.

The San Andreas fault and the trajectories of the two rock masses it separates also provide an excellent model for envisioning *how* many continental margins interact. One of the great discoveries of the continental drift model is that continents not only slide across the surface of the globe like lugubrious bumper cars, alternately smashing into and fleeing other continental entities, but also *grow* through plate tectonic processes. Small slivers of other land masses—islands and pieces of other continents—these lithological flotsam drift ashore and meld on the edges of larger continents through the long eons of time. And in the process, they inexorably enlarge various continental bodies.

North American is no exception to this steady continental enlargement. Starting as a central core of ancient rocks such as gneiss, granite, and schist several billion years ago, North America grew as successive waves of small lithic immigrants either accreted onto its coasts or became cemented in place due to the collision of North America with some other continent. Along the eastern seaboard of North America, the Appalachian Mountains swelled upward and outward, first by mountain formation inland of active subduction zones, later by the collision of North America with Europe and Africa.

Mountains also grew in western North America, but differently than in the east. Subduction took place, creating high volcanoes. But many more, smaller pieces of drifting crust collided and became accreted onto our wandering continent. Exactly how and *when* such smaller pieces accreted onto the western coast of North America has been the source of enduring dispute.

A territorial dispute

Perhaps no block of this itinerant real estate has created so much controversy as the region encompassing Vancouver Island and Western British Columbia. The conflict is over whether this giant region was formed through the collision of one or more subcontinents smashing into the larger North America coast and thus is the first and still-enduring example of what is now known as *accretion tectonics*. One side believes that Western British Columbia has always been in its present-day position. The other believes that all of this vast region drifted northward during the Age of Dinosaurs, finally to collide with the then-westernmost coastline of North America, producing the Coast Ranges of Canada and the North Cascades of Washington State.

Plate tectonics gave us the overall view of how the largest pieces of the earth's crust, the continents and ocean basins, have interacted through time. *Accretional* tectonics tells us about continental assembly and how coastal mountain belts form. The controversy often arises from the simple fact that mountains are the most complex geological units on the planet. Because mountains are composed of rocks thrust and shattered, folded and heated, pressurized and flung high into the sky, they are usually impervious to simple structural analysis of their component parts. To put it more simply, the characteristic trauma of mountain building destroys the evidence about their formation.

Mountains are formed by several processes, including compression, extension, and the eruption of volcanoes forming in response to subduction. But when these processes are melded with accretional events—such as an island the size of Vancouver Island colliding with an already-forming coastal mountain range—things become much more complex. Today, the most researched aspects of continental drift are related to the importance of such collisions between continents and the smallest crustal fragments accreting onto continental edges. These ephemeral and active bits of land are called *suspect terranes*, because more often than not their formation in this manner is hypothetical—and usually controversial as well.

John McPhee, perhaps the greatest living writer on things geological, has visited the topic (as well as the real estate) of suspect terranes in several of his books. They are wonderful subjects: regions where large pieces of mountain chains are seemingly exotic, perhaps far-wanderers coalesced into more banal country rock. They come in various shapes and sizes, and to discuss them we need some specialized terminology.

The first important pieces of such puzzles is *microplates*. Most plates (the *plate* of plate tectonics) are enormous affairs, covering significant portions of the earth's crust. It is one of the great geological mysteries *why* there should be so few plates making up the surface of the earth. North America, for instance, though gigantic in itself, is only one piece of a plate that also includes half the Atlantic Ocean. The plates that make up the preponderance of the vast Pacific Ocean are even larger. But interspersed among these behemoths are smaller versions, the microplates. When microplates and larger plates interact, quite complex geological events can occur. Take, for example, the region around Indonesia. It rests on a microplate squeezed between three behemoths: the India–Australia plate, the Pacific Plate, and the Eurasian plate. Australia at the present time is moving forward, slowly squeezing all of the myriad island around Indonesia *into* Asia in the process. The result is geological chaos: The collision is smashing volcanic island chains, sediment-filled basins, pieces of continental margins, oceanic seamounts, and young oceanic crust into the same geological blender. Each of these units is composed of its own rock type, and as they squish together and ooze over one another, a most complex assemblage is formed. Some pieces are raised upward, others are depressed and run over; mountains form, mountains are destroyed. Pity the poor geologists of 50 million years from

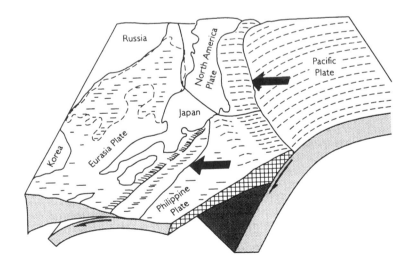

Subduction zone along the descending Pacific Plate.

now (unless they have a *real* time machine) trying to deduce the sequence of events. But just such chaos appears to be what formed the west coast of North America 50 to 100 million years ago, and pity poor *us* now trying to work it out.

A second key unit is the *terrane* (more properly a "tectonostratigraphic terrane," but let's keep it simple). Terranes are collections of rock showing a particular stratigraphy that is separated from other rock types by faults, or breaks in the rock. Microplates become terranes when they collide with other rocks and are fused onto continents.

Terranes (and the suspect ones among them) have several disparate origins. They can be (1) fragments of continents, long since split off, composed of old sedimentary rock and perhaps granites and old metamorphic rocks; (2) fragments of continental margins, which are usually composed of sediments shed from a continental margin); (3) fragments of volcanic arcs, made up of accumulated lava and sediment; or (4) fragments of ocean basins, and thus made up of basalt.

Terranes may begin their existence by being "born" in some oceanic setting (such as through the accumulation of lava in a submarine setting), or they may form as they are "calved" off of some already existing land mass, through a process known as rifting. In each case the accumulated rock is buoyant relative to oceanic crust, so it "floats" on heavier crust. Its fate, sooner or later, will be to collide with some other buoyant crust. Such collisions are the process through which continents often grow larger, for more often than not, the colliding pieces accrete onto the larger land mass. But how can we deduce the antiquity of such an event, and how can we tell what the colliding pieces looked like, were made up of, or came from before the collision? Stratigraphy is one powerful tool, as we saw in our discussion about the first recognition of the Gondwana supercontinent. But an even more useful tool comes from paleomagnetics.

There has long been an uneasy alliance between the two great pillars of the earth sciences: the branches known as geology and geophysics. In the simplest sense, geology addresses history—*when* the observable features on and in the earth formed—whereas geophysics deals mainly with the physical forces that did the forming—the *how* and *why*. Geology is thus more concerned with time and geophysics with process, although both fields borrow from the other, and the distinctions between them are anything but clear-cut. Nevertheless, although they work most often in concert, the fields of geology and geophysics have sometimes found themselves on opposite sides of fierce scientific debates. This is especially true for two of the most important questions confronted by the geological sciences: How old is the earth? Do continents drift?

With regard to the age of the earth, the father of geophysics, the great Lord Kelvin, and his followers were certainly wrong, and these errors led many scientists astray for decades. The geophysicists made a similar, monumental blunder when it was originally proposed, in the first half of the twentieth century, that continents had drifted. The geophysicists simply could not conceive of a way in which such huge blocks of matter could sail across wide ocean basins. Geologists, especially geologists who studied the Southern hemisphere, such as Wegener and du Toit, amassed tomes of stratigraphic and other evidence: the fit of the continents, the presence of similar fossils, and the similarity of rock types or succession on continents now widely separated. Today, these lines of evidence appear as incontrovertible as they did to the ardent and frustrated geologists of a half-century and more ago. Yet physics and its branches, of which geophysics is one, have long assumed the mantle of scientific supremacy. The geophysicists held firm in their vehement opposition to continental drift and, as in their stance concerning the age of rocks, retarded scientific progress for decades.

This chastisement is offered as a prologue to irony: The shoe has traded feet. Now, in many regards, it is *geology* that has accumulated blinders and geophysics that frets. The subject of concern is still within the province of plate tectonics, but one subdivision thereof. The controversy involves suspect terranes and is focused on the formation and history of western North America.

Perhaps the greatest single tool whose use eventually convinced geophysicists that the crust of the earth is at heart a wandering entity came from the study of rock magnetism. As we saw in Chapter 3, any igneous or sedimentary rock solidifying for the first time usually contains tiny magnets that can yield highly accurate information about both polarity and geomagnetic field directions (which are themselves dictated by latitude). These two bits of information became crucial lines of evidence of continental drift. In areas where new sea floor is being created at the mid-ocean spreading centers, a series of linear magnetic stripes of opposing polarity have been measured by ships and planes towing sensitive air- or waterborne magnetometers. Second, the common application of magnetic coring from continental rocks showed that different continents seemed to exhibit very different directions of the ancient magnetic poles. Successively older rocks in North America, for instance, seemed to suggest that the position of the earth's magnetic pole has been migrating. Of course, it was not the earth's magnetic dipole positions that were wandering but *the continents themselves*. By tracing them back through time, via paleomagnetism, detailed histories of drift could be obtained.

Such was the up-side of paleomagnetic analyses. The down-side is that rock units that have been reheated gives spurious results. Unfortunately, no easily observable sign alerts the investigator that this has happened. The magnetic resetting of mineral grains occurs most frequently in those parts of the earth where regional heating and high pressure are most intense—in mountainous regions. Because these are the regions that are usually of greatest interest for deciphering a continent's history, paleomagnetic work rarely yields a simple interpretation. The mountainous regions harbor the various slivers and shards of crust formed by collisions of continents with other rock units, and it is in the mountains that paleomagnetic studies are most urgently needed—and most perilous.

Within a decade of victorious overthrow of all preceding theories, global tectonics and continental drift theory had revealed the path of North America's drift over the last 500 million years (although many details still remain to be discovered). But much more mysterious is the history not of North America's *drift* but of its tectonic formation, and no region remains more obscure than its west coast. Stretching from Alaska to southern Mexico, virtually the entire western margin of North America is mountainous. How did these mountains form? Were they the product of subduction only, where ocean crust dives under a continental margin, melting (and creating a line of volcanoes on the surface) in the process? Clearly there were many volcanoes along the west coast of North America, and as the movements of the oceanic crust was deciphered in the early 1960s, it became clear that many of the mountains of North America were indeed produced by subduction. The high Cascade volcanoes of the Pacific Northwest are an excellent example. Equally clear, however, is the fact that this type of plate-to-plate contact, creating subduction zones, was not the only type of plate interaction going on. Other processes of mountain building have been taking place as well. Perhaps nowhere are these latter types of plate interaction better studied than in California.

California is earthquake country. Earthquakes occur when rocks move. Most of California's earthquakes can be attributed to the type of plate-to-plate motion known as strike slip. This type of movement, wherein the edges of plates scrape by each other, is causing the westernmost portions of the Golden State to move northward. This motion will continue until the various fragments, carrying San Francisco and Los Angeles on their backs, run into something, such as another part of the west coast of North America.

It now seems probable that much of western North America was created by collisions with smaller, subcontinent- or island-sized land masses, forming mountains in the process. We are witnessing the start of one such voyage; its destination is still unknown. San Francisco and the long sliver of land it sits on will, over the next few millions of years, leave North America's embrace to become a long borderland west of the continent, thereby creating an inland sea. Sometime in the far future, it will probably collide with North America again. This seems to have been the tectonic style for westernmost North America for several hundreds of millions of years: Land masses large and small have been moving northward and colliding with western North America, becoming part of the continents in the process, or sometimes continuing their northward journey. Geologists have deciphered this history not only by studying the San Andres fault complex, but also, where possible, by conducting paleomagnetic analyses for various bits of North America.

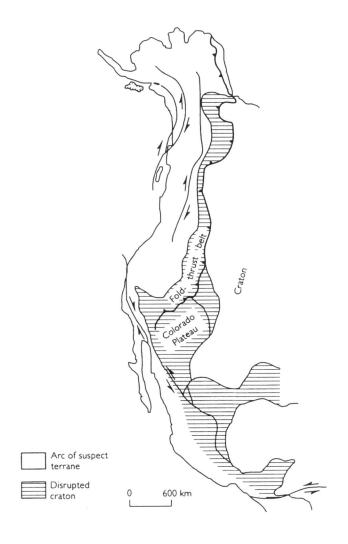

Locations of craton, disrupted craton, and suspect terranes in Western North America. Major fault zones and respective movements along these faults also illustrated.

For more than a century, deciphering the tangled, complicated mixture of rocks that make up western North America has been the career choice (or fate) of hundreds of the best earth scientists. They have needed to be smart: The variety of assemblage and style seemed incomprehensible for a long time. The emergent theory of plate tectonics—and especially the new tool of terrane analysis using paleomagnetism—seemed at last to offer a way out of this morass. The observed direction of northward tectonic movement in the present-day California coast region supplied a key clue to how the coast had formed. Earth scientists began to wonder if much of the west coast of North America had not formed far to the south, as small islands or subcontinents, and then become accreted onto the continent. In the early 1970s, teams from Canada and the United States began sampling various rock bodies for paleomagnetism in order to test such ideas. These data eventually became the primary weapons in a war pitting the geophysicists, armed with their paleomagnetic information, against the many geologists who believed that plate tectonic models requiring large-scale transport of rock bodies over distances as much as several thousand miles were not necessary to explain the complicated geology found in western North America.

The opening shot was fired in 1971, when Robert Tessier and Merl Beck of Western Washington State University published paleomagnetic data from the volcanic rocks making up Mt. Stuart, a forbiddingly high peak in the North Cascades of Washington State. Compared with the largest of the Cascade mountains, such as Mt. Rainier, or the nearby Mt. Baker (which is still an active and exceedingly dangerous volcano), Mt. Stuart is far more ancient, and it is "extinct." It has been dated repeatedly, using sophisticated radiometric dating techniques, and all measures yield a Middle Cretaceous age of about 100 million years ago. The dispute about Mt. Stuart is not about its age but about where it was formed.

The pioneering work by Tessier and Beck resulted in a startling discovery: They found that Mt. Stuart was made up of rocks seemingly solidified at a latitude far to the south of its present position. If the paleomagnetic results were correct, then the gurgling, bubbling, molten magma that 100 million years ago solidified into what is now Mt. Stuart did so somewhere off what is now Southern California or Mexico. This discovery shook the geological establishment to its core.

Soon afterward, other paleomagnetic findings suggested that not just Mt. Stuart, but also other pieces of northwestern Washington and southern British Columbia, journeyed from lower latitudes. Each of these rock bodies had to have traveled between 2,000 and 3000 kilometers from the south. But here was more than simple transport. The rock bodies were also rotated as

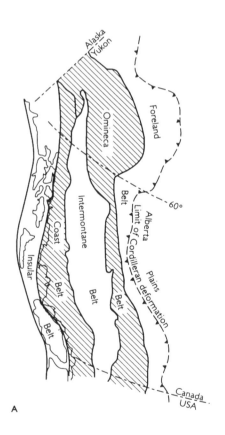

A

A) Morphogeological belts of Western Canada.

much as 70 degrees—as though a giant had grabbed the land and twisted it
to the right.

But was the paleomagnetists' interpretation correct? Could the various
volcanic rock bodies that they examined *not* have been transported and yet
have yielded such results? "Yes!" answered many geologists. Exactly these results would be expected if the various sampled rocks had been tilted, rather
than transported.

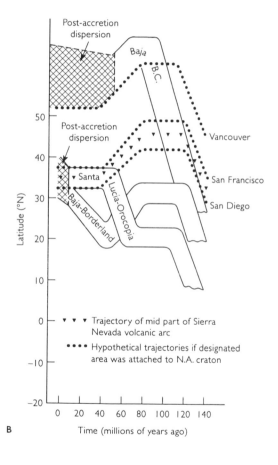

B) Travel paths of various regions in Western North America relative
to the craton (stable continental regions).

It is much easier to tilt a rock than to transport it. Tilting takes place all the time: just look at any range of mountains and you will find rocks in every conceivable orientation, including turned upside down. The paleomagnetic interpretations relied on the ability to recognize *original horizontally* of the sampled rocks. This was a key argument leveled by critics of paleomagnetic studies: The rocks sampled were not horizontal, and therefore the results were spurious.

Thus the controversy started, pitting "drifters," or those who believed that the various tectonic pieces now making up the Western Cordillera had assembled by continental drift, against "fixists," who believed that no such assemblage of exotic terranes had taken place. The fixists did not doubt the validity of continental drift; they just believed that many continental drift theories as applied to the western United States and Canada were simply figments of overwrought geophysical imaginations. Besides, argued the fixists, Mt. Stuart and the North Cascades are not located on the coast but are miles inland. If they had been transported, then surely all rocks to the west of them had to have been transported as well, and there was no apparent evidence of that. Then, in 1977, a study defining and identifying the greatest suspect terrane of all was published, and this significantly changed the debate.

A prime suspect

The definition of the giant exotic terrane known called *Wrangellia* (after the Wrangell mountains of Alaska) was for several reasons a watershed event in plate tectonic study. First, the study was published by three senior and highly respected geologists employed by the United States Geologic Survey: David Jones, Norm Silberling, and John Hillhouse. Second, until that time most exotic terranes were hypothesized to be small land masses—island-sized at best. The Wrangellia Terrane as first envisioned was enormous: a relatively thin strip of land running north and south for many hundreds of kilometers. Finally, the means used to identify this ancient land mass and its drift history provided a methodological example for most subsequent examinations of exotic terranes. In the Wrangellia paper, both geological evidence and paleo-magnetic evidence were used, rather than only one or the other.

The Wrangellia terrane was defined in terms of geological criteria. It is composed of Triassic-aged basalt that appears to have been formed deep under the sea. More than 200 million years ago, enormous outpourings of basalt quite similar in texture and composition to those making up the Hawaiian Islands erupted from great submarine fissures. This basalt continued to burp out for many millions of years and eventually accumulated into a mass of hardened lava many thousands of feet thick. Finally, in late Triassic

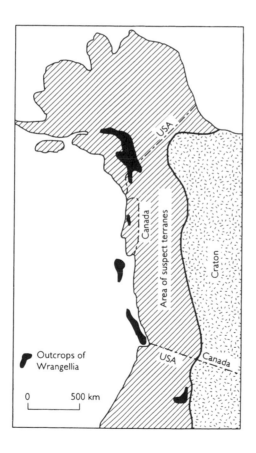

Present distribution of Wrangellia terrane.

time, the fissure closed; no new lava emerged. By this time the lava had formed into a long, linear land mass. Limestone began to accumulate after the volcanic eruptions ceased, and the underwater environments where these limestone deposits formed were colonized by early Mesozoic creatures, including corals and many species of ammonites.

Wrangellia was enormously long. It extended from Vancouver Island as far as north Alaska. It may have been even longer: Isolated scraps of rock in the Hell's Canyon region of Oregon and Idaho show the same telltale sequence of rocks used in the identification of Wrangellia.

Both paleontologists and paleomagnetists found that Wrangellia was even more far-traveled than Mt. Stuart. The origin of the Wrangellia basalt was somewhere in the *southern* hemisphere. The entire rock body had then been carried northward by continental drift, eventually to crash into, and fuse onto, the western coast of North America.

When did this titanic collision take place? In the early 1980s, paleomagnetists made extensive studies of Eocene-aged rocks at one site near Mt. Stuart and at another to the east of Vancouver Island. In both regions, a great thickness of 50-million-year-old sandstone accumulated in basins near a mountainous region of high relief. These sedimentary rocks contain palms and other fossilized tropical plants, which suggests that they too may have come from more southerly regions. But the world of the Eocene appears to have been warmer and wetter than that of today—a place where tropical forests could have existed even at the latitude of what are now Washington State and southern British Columbia. The paleomagnetic results confirmed this; none of the Eocene-aged sandstone showed any evidence of drift. These studies place an upper time limit on the collisions of Wrangellia with North America: 50 million years ago.

The idea that this long, thin terrane had smacked into North America sometime in the late Mesozoic had just filtered down into textbooks when, in the mid 1980s, the whole concept of Wrangellia changed. It began to grow in the eyes of some geophysicists. The prime mover in this was Ted Irving of the Geological Survey of Canada. Irving, a paleomagnetist, was stationed in Victoria and thus had ready access to British Columbia rocks.

His studies began to convince him that far more than Vancouver Island and the Queen Charlotte Islands (the two prime pieces of the Wrangellia terrane in Canada) had moved up the coast. Irving began to wonder if not just this terrane, but something larger, had traveled up. Soon Wrangellia received a new name: the Insular Superterrane. It now encompassed not only Vancouver and the Queen Charlottes but all of the coastal mountains of British Columbia as well. The vision now was of a *huge* hunk of real estate—a subcontinent something like India in size, but longer and skinnier—smashing and accreting onto the coastline. Irving coined a new name for this superterrane, calling it Baja British Columbia after Baja California, the latitude from which he believed it had come, so long ago. The entire idea became known as the *Baja British Columbia hypothesis*.

Irving was no stranger to controversy. As a graduate student at Cambridge University in the 1950s, he had been one of the first of his generation to embrace paleomagnetics as a tool for studying past continental positions. His studies, which were conducted just before the emergence of continental drift and plate tectonics as an accepted theory, clearly showed that either the continents were drifting or the poles of the ancient earth had bad tendency to wander. Irving refused to accept the latter possibility and thus embraced the idea that the continents had drifted. Unfortunately, Cambridge University was not so open-minded and refused to award him his hard-earned PhD. Irving was forced to start again at another university. Only much later did a very sheepish Cambridge award Irving an honorary PhD.

As the 1980s progressed, even this enormous land mass grew larger in the conception of the geophysicists. Now, not only the coastal mountains and Vancouver Island in British Columbia (the Insular Superterrane) were thought to have drifted; the Intermontane region—the entire suite of gigantic mountains that make up most of British Columbia—was added into the mix as well. By the late 1980s, geologists from the University of Washington (Darrel Cowan and his students Paul Umhoefer, Mark Brandon, and John Garver) had refined this model. They put it on a far more rigorous footing by conducting kinematic studies showing how the plate movement could have produced this event.

According to all studies, the southern limit of Baja British Columbia is found in the San Juan Islands of northern Washington state. In fact, Sucia Island may mark one bit of this southern boundary. The rocks making up the fossiliferous portions of Sucia, as well as other parts of the Upper Cretaceous Nanaimo Group of the Vancouver Island region, suddenly became critical elements if this idea of a drifting subcontinent smashing into North America was to be tested. Mt. Stuart, 100 million years old, seemingly obtained its magnetic signal somewhere in the latitude of Mexico. The Eocene rocks of northern Washington seem to have originated at a latitude similar to that of their present location. The rocks of intermediate age, the fossiliferous rocks of the Nanaimo Group—including Sucia Island—would seem to hold the key. Two teams of American and Canadian scientists, led by Merle Beck and Ted Irving, quickly began to sample the rocks of the Vancouver Island region. Here, surely, would be the key to unraveling—confirming or refuting—the mystery of Baja British Columbia. Did it exist at all? How much drift took place? To their dismay, they found only nonsensical results. The problem, again, was reheating.

Magnetic crystals locked in rocks provide the compasses that reveal ancient pole positions. However, if the rocks in which they sit are reheated, the force lines of the magnetic field change orientation at the time of reheating. All previous information is erased, or overprinted. Sometime in the early Tertiary Period, perhaps 50 million years ago, large portions of Vancouver Island were reheated. The cause of this heating is poorly understood. Perhaps the island ran over a hot spot, or perhaps regional volcanism simply heated larger parts of the region. The result, in any case, was to scramble the precious magnetic information locked in the rocks— information that could have been the decisive test of drift history. Because of this, it was soon widely thought that no reliable magnetic signal would ever be recovered from the Vancouver Island region. The paleomagnetic results from my drilling on Hornby Island in 1986 only seemed to confirm this view. But as so often happens in science, a chance event would alter this perception.

Baja California

Late November in Seattle is a bitter time. Summer is long gone, only a distant memory. With November come the Pacific gales, bringing unrelenting rainfall. Days pass with no glimpse of the sun, and in the high latitude of the Pacific Northwest the November days are short. Darkness comes early, and the spirit is challenged. No tourists visit Seattle in November, and the number of suicides skyrockets. It is a great month to escape southward, to sunnier climes and ammonite-rich rocks. No wonder November has been chosen as the prime time to study geology in Baja California.

There is no easy way to get to Baja. My own route always passes through Pasadena, home to the California Institute of Technology. On my first-ever trip to this fabled institution, my plane landed in Burbank in the dark. Then I made my way to Pasadena by shuttle, finally to be deposited at the front door of an impressive stucco building: the Athenaeum, Cal Tech's faculty club.

Faculty clubs are normally somewhat cheesy affairs tucked away in some forgotten corner of a university campus and frequented by the poorly dressed army of university faculty. Yet this place was elegant, old-money Californian. The upper stories of the building formed a comfortable hotel. Downstairs was a huge living room, walls covered with portraits that constituted a veritable *Who's Who* of American science in this century. Rich, thick carpets leading to a large dining room muted murmurings from well-dressed patrons (most of whom, I later learned, were not faculty members but wealthy Californians willing to shell out large membership fees to rub elbows with the distinguished faculty). Dressed as I was in field gear and hiking boots, I looked like a man in a Hallowe'en costume, and the other patrons looked at me as if I were a Martian.

I had traveled to Cal Tech as the first leg of an expedition to Cretaceous-aged exposures found along the western coastline of Baja California. My host for this trip was a man I knew by reputation but had never met, the paleomagnetist Dr. Joseph Kirschvink. After settling into my room, I set out to find Kirschvink. I was told that he was currently speaking in one of the largest of the university's lecture halls. It seemed odd to me: past 8 PM and still teaching. I made may way across the gardens and walkways of the Cal Tech campus to the lecture hall in question. I came to the appointed door, entered, and saw a sea of humanity, perhaps 300 people. It was also clear that those packing the lecture hall were not students; they were too well dressed, too old. I found a seat and focused on the speaker: a somewhat diminutive man from a distance, a rather high speaking voice—and the most extraordinary subject. Kirschvink was telling his audience that there is a *sixth* human sense in addition to the well-known five. All of us, he said, have an innate sense of *direction*, produced by small crystals of magnetite found within virtually every brain cell, a discovery made largely by Kirschvink himself. This was my introduction to my new scientific partner.

I had been invited to join a very famous field trip. Every year since the early 1980s, Kirschvink has taken several loads of students (and the occasional lucky outsider such as myself) to Baja California to study the local geology. I had heard about this trip from various colleagues for several years and by sheer good fortune had managed to get myself invited, not knowing at the time that our road to Baja would eventually take us back to Vancouver Island.

We assembled the next morning to load the vans, bought huge amounts of groceries, and set out. The Cal Tech Baja trips entail endless driving. Three hours from Pasadena to the border at Tijuana, then another two or so to Ensenada, and then south on Highway 1. An hour south of Ensenada we turned west into the hills on a track that could only euphemistically be called a road, over rugged countryside and washboards, the entire van now filled with dust.

We finally arrived, in early evening, at our camp site. Tents were pitched in the dark under starry skies, but a howling wind from the nearby sea drove us into sleeping bags early. Morning, however, brought epiphany. Emerging from my REI special I found that our tents sat atop a rugged seacoast cliff, near a tall white lighthouse. A dirt track led down to the sea, where green shale made up low coastal cliffs stretching to the horizon. I descended to the beach and its rocky cliffs, crossing the intertidal rocks gloaming under a retreating tide. I watched pelicans wheel overhead. I noted the remains of rock lobsters and tropical snail shells, feeling the glow of salty air and a warm lapping sea—in a geological setting made up of rocks and fossils so familiar as to be bizarre in this near-tropical context. The rocks were old friends from my earnest boyhood days on Orcas and Sucia Islands. And it wasn't just the rocks. Far more telling were the *fossils*, old friends too, ammonites and snails familiar from the Vancouver Island region, yet sitting here entombed in this gray green matrix on a *Mexican* not Canadian coast, announcing their names with a Spanish accent: *Baculites, Hoplitoplacenticeras, Desmophyllites, Inoceramus*. Perhaps it was coincidence that these rocks and those of Sucia and Vancouver Islands shared fossils as well as lithology. But perhaps not.

Joe Kirschvink had studied the paleomagnetic signals of these rocks over the years to ascertain their ancient latitude. His questions were thus fundamentally different from the research questions I had been examining via paleomagnetics. Joe was interested in ancient continental positions; my work, up to this point, had been concerned only with time as calibrated by the geomagnetic polarity time scale—the record of geomagnetic reversals. Joe's work on this beach demonstrated that the paleolatitude of Baja California some 75 to 80 million years ago was nearly the same as it is now: between 20 and 25 degrees north latitude. But what of Vancouver Island? Was the Baja British Columbia hypothesis correct?

We spent several days touring various outcrops along the coast, spending little time on any given outcrop. Nights were filled with food, Corona, nips of Tequila, and much conversation. On this trip, Joe and I launched a friendship and a scientific collaboration as well. Neither of us much liked the possibility that the Vancouver Island rocks were completely reheated, and we discussed areas that had not yet been sampled, regions that might contain an original, Cretaceous-aged paleomagnetic record that could give reliable paleolatitude for the Vancouver Island region at a time when dinosaurs still ruled the earth. It was on this first Baja trip that we began to make plans to mount yet another assault on the paleomagnetics of Vancouver Island, this time armed with better sampling techniques and a far better paleomagnetic laboratory to analyze the cores. It took several years and a grant from the

National Science Foundation to bring these plans to fruition. On the Canadian expedition, it was Joe's turn to fly.

Baja British Columbia

We arrived on Hornby Island in early summer of 1995: Joe Kirschvink, Jose Hurtado (a student assistant from Cal Tech), and me. We walked the beach at the lowest of tides, the sun glancing off an occasional ammonite shell still held tightly in the sediment. The ammonites, it turned out, gave us our best hope of finally getting a real paleomagnetic position of these ancient sediments. The shells and the shapes of the fossils were our clue. Temperature is the key to paleomagnetic analyses; if the sediments were heated too much, they lost their magnetic signal. But our ammonite shells also reacted to high temperature: Their pearly shells retained luster only in the absence of heat. Any regional heating would recrystallize these shells to a dull tan or even black. Because of the pristine nature of the ammonites from Hornby and from a few other islands of the Vancouver Island region, we were confident that a reliable paleomagnetic signal could be found. The key to success this time on Hornby would be avoiding rocks that appeared to be oxidized in any way or did not have a fresh surface. Joe's solution was to excavate all surface rock deeply away before drilling, to ensure fresh rock.

For several days the three of us worked in exquisite weather, camping nearby and sampling island cuisine after work. With more than 60 cores obtained, we moved north to one of the largest and most obscure of the islands in this region, a place called Texada Island. Here too we sampled the strata, this time with much more difficulty, for the timber companies had devastated this island. Whereas on Hornby the fossiliferous rock is exposed on a seacoast, on Texada the sampling locality was an overgrown creek that years ago had been clear-cut. The timber company that did the cutting on this creek never cleaned up the snags, and the creek was an overgrown mass of felled dead wood and colonizing brambles. Our prime locality could be reached only by crawling half a mile over the snags carrying our heavy gear, and it was relief when we finally finished this work without anyone breaking an ankle or worse. The field sampling was finished. The next phase occurred at Cal Tech, deep in the basement of the geology building, where Joe maintains his paleomagnetics lab.

A magnetometer is an odd device cooled by liquid nitrogen. The small cores we had extracted from the rocks of Hornby and Texada Islands were sent downward into the freezing heart of this machine like Han Solo being turned into a wall hanging at the conclusion of *The Empire Strikes Back*. In go the cores, out come the cores, and numbers flash across a digital readout screen. Each core is repeatedly analyzed, and then begins the long process of removing spurious magnetic overprints. The cores are heated in incremental steps: first 100°C, then 150°C, and on up in 50°C increments. After each heating, which takes about an hour in the oven, the core is analyzed again. Slowly the obscuring overprints of heat are stripped out of the cores; gradually an original signal comes into view; and if all goes well, after many hours, each core provides one datum point. Jose Hurtado had done much of this work for the Hornby and Texada cores, and I had come to see the final results. Joe had taken the data and analyzed them using a statistical method he had devised more than a decade earlier. On page after page of printout he now held the end of a long journey: the mean latitude of both Hornby and Texada Islands was 25 degrees north latitude at the time when their rocks were deposited, late in the Age of Dinosaurs, about 80 to 70 million years ago. Both islands show the same latitude as Baja California today. There are great moments in life. For me, this was one.

I wrote an account and submitted it in all our names to *Science* magazine. Then we waited several anxious months for the reviews to come in. *Science* is the most difficult of all scientific journals to be accepted in; its editors require that the scientific problem being examined be of general interest to *all* scientists and they require that two external referees give the manuscript a ringing endorsement. On the day the letter arrived, I ripped open the envelope announcing the official decision. Despair! One of the reviewers had flatly dismissed our results as "spurious." He decreed that we had simply measured effects of a disputed phenomenon called "inclination shallowing." It was clear where this review had come from. *Science* had sent our manuscript to die-hard critics of this particular terrane drift—two men who do not accept the validity of the Baja British Columbia hypothesis.

Inclination shallowing is a phenomenon wherein compaction of sediment, caused by the slow but inexorable accumulation of sedimentary beds one atop another, eventually flattens underlying sediment to the point where the magnetic signal "shallows," or gives evidence of a lower latitude than was originally present in a bed. For example, some sedimentary bed might have been deposited in latitude 45 north. However, the gradual squashing of this sediment over time may slightly flatten out the small magnetite grains such that the paleomagnetic signal measured might be 34 or 35 degrees. These two reviewers had dismissed every Mesozoic paleomagnetic of tropical latitudes in now-northerly localities as either due to inclination shallowing or coming from studies where the original horizontal attitude of the rocks could not be ascertained. Our paper was rejected.

If I was crestfallen, Joe was outraged. Our study was *not* contaminated by compaction shallowing, he announced to me over the phone soon after our rejection, for a simple reason: The ammonites, so abundant in the sediments from Hornby and Texada, were not compressed. If compaction shallowing were taking place, it would affect not only the magnetite but the fossils as well: The fossils would be flattened. Clearly they were not, so inclination shallowing could not have taken place.

We communicated this to the editors in an appeal, and after more reviews the paper was accepted. Published on September 12, 1997, it presented a data set that gave more credibility to those who maintain that the "suspect terrane" known as Baja British Columbia is less suspect than most.

Who—or what—was watching, so many millions of years ago, when the entire block of what is now British Columbia crashed into North America to be welded in place? Magnetometers are the time machines that have revealed to us this surprising event in earth history, when ancient Mexico cast off its moorings and sailed north. Canada is a far richer place as a result. But will the Mexican government, so keen on repatriation, demand its territory back?

5

Ancient Environments
and the Level of the Sea

Sedimentary rocks continuously change in appearance as we follow them either up or down into time. The stacked layers become more fine- or coarse-grained, and they change from one type of sedimentary rock to another. For rocks deposited beneath the sea, it seems that the depth of deposition is the prime factor in dictating these changes. Water depth and changes of water depth (which in marine environments is termed *sea level change*) have played a large part in determining what type of sedimentary rock is deposited in any given underwater locale—and thus have played a major role in fashioning the appearance of our planet's sedimentary rock cover. But sea level change has been blamed for more than simply influencing the fate of ancient sand grains (and thus complicating geologists' lives, as they grapple with describing sedimentary rocks). It has also been implicated in the greatest mass murders to have occurred on this planet. Sea level change is a prime suspect in the periods of mass death, called mass extinctions, that shaped out planet's biota. The way geologists have deciphered sea level change is thus a potent type of time machine for planetary detectives on the global homicide detail.

Trekking through time

The study of sea level change is part of the discipline of sedimentology, the study of the way sedimentary or layered rocks form. This branch of geology gives us the tools necessary to reconstruct the geography of ancient lands. Sedimentary rocks can be seen forming today in a variety of environments on land and in water, such as in deserts, lakes, rivers, and oceans; and they surely did so in similar fashion in the past. This belief forms the core of the Principle of Uniformitarianism, which holds that the present is the key to the past. For instance, the way in which sand grains accumulate into sedimentary beds on a modern-day beach or riverbed is probably the way such processes occurred on ancient beaches or riverbeds. Time may change, but the laws of physics, which control the accumulation of sedimentary rocks, do not. The principle of uniformitarianism is most powerfully used in the field of sedimentology, where it enables us to reconstruct past environments. In like manner, we can often reconstruct and deduce the "life habits" of ancient organisms by studying how modern forms of similar species aggregate and live.

Each sedimentary environment, be it a riverbed or desert floor, produces a specific type of sedimentary rock, for the physical forces operating in each particular environment frequently dictate how the resulting assemblage of sedimentary rocks will look. Deserts, for instance, accumulate sandstone amid wind-blown sand dunes, which more often than not leave a record of their existence in the rock itself, in the form of swirls and irregularities in bedding known as cross beds. Any such marks in sedimentary rocks are known as sedimentary structures, and these often provide the most specific clues to ancient sedimentary environments. The sedimentary rocks forming in a deep ocean look quite different from desert deposits, because they have accumulated in much quieter, lower-energy environments and consequently show different grain sizes and sedimentary structure than do desert deposits. Any fossil types present are also important clues to reconstructing the environments in which a particular sedimentary rock accumulates. Organisms are restricted to very specific types of environments and, when preserved as fossils, are among the best of all sources of information about what the "place" was like where a given assemblage of sedimentary rocks accumulated. When all of these types of evidence are considered together, they can yield a detailed understanding of an ancient environment.

The sedimentary rocks on Sucia Island, for example, can be analyzed in this way, by combining the evidence gleaned from grain size, sedimentary structures, and the enclosed fossils. To make such an analysis, we need to "walk the section," beginning amid the lowest (and hence oldest) of the island's sedimentary rocks and finishing amid the highest, or youngest. If Sucia's sediments retained their original horizontality, we would need a huge stepladder to complete such a task. But Sucia's rocks have been contorted in such a way as to make our job much easier.

The sedimentary rocks of Sucia Island have been subjected to enormous compressional forces that have deformed the region such that the island, as seen from the air, now describes a series of gigantic nested horseshoes tens of miles across. We can see how this occurred by "performing" the following "thought experiment." Imagine that the ancient sedimentary beds of Sucia Island are represented by a telephone book, each page representing one layer of strata. The book lying on a flat surface represents the strata of Sucia prior to deformation, each page horizontal and parallel to every other. However, the Sucia strata were not to be left in such peace. Imagine that the two sides of the book are squeezed, causing it to assume the shape of a half tube, with the spine of the book and the pages on the open side pointing up. If we now take this deformed book and jam it into the surface it sits on at a 60-degree angle, we have roughly reduplicated the forces that the Sucia Island strata have undergone. The resulting curvature is sure evidence that the

Part of the sedimentary section on the south coast of Sucia Island. The lowest (oldest) beds are at the bottom of the photo, the highest (youngest) at the top.

once-flat-lying sedimentary rocks have been involved in a regional compressional episode of the sort that can produce mountains. Millions of years ago, the flat layer cake of strata making up Sucia was slowly pushed against an unyielding wall of mountains to the east. The layers first bent, then buckled and broke. All such effects are manifest on Sucia as tilted strata and faults. The originally flat layers dip at an angle of nearly 60 degrees, making the complete book of strata that is now Sucia Island easy for us to read. Simply by strolling along a beach of the island, we can walk through time—through the entire pile of sedimentary beds.

The oldest sedimentary rocks that make up Sucia Island are located on its southern margin. Let us start out there on a sparkling May morning, warm sun prickling our still wintry skin, and walk the intertidal, one eye watching our step on the beach strewn with boulders and driftwood, the other watching the ever-changing cliffside rocks. We will need a low tide, because there are several landslides that would otherwise ban our way along the path we must follow on Sucia's southern coast.

The most ancient strata on the island are conglomerates, the type of rock made up of larger rock particles, such as gravel and cobbles. These conglomerates on Sucia are a chaotic and jumbled pile of rounded rocks, a few as large as a bowling ball. In some areas the conglomerates contain beautiful white quartz; in other places a more variegated rock assemblage is visible. The various rocks in this basal conglomerate are sure clues to the nature of the land area that was nearby when these rocks were deposited, for they are so large that they could not have been transported far from their original mountain sources. We can surmise that these cobbles were brought to the edge of a sea by rivers tumbling down from nearby coastal mountains; in this conglomerate we see bits and pieces of these ancient mountains. They give us our first glimpse of ancient Sucia Island—it was near mountains.

The aggregate thickness of Sucia's basal conglomerate is perhaps 100 feet, but because the beds are tilted on their side at a 60-degree angle, we rapidly pass through this thickness of rock as we walk along. The rocks are exposed both as a low cliff lining the beach and as a bench along the beach itself. There is very little bedding or any other type of sedimentary structure visible in the conglomerates. They make up an imposing coastline. As we continue our walk, we eventually come to a transition, where the conglomerate is overlain by another sediment type, one dominated by sandstone. This gradual transition is best seen in the rocky sides of the cliffs lining the beach. Over several meters of thickness, the rocks in this conglomerate grow finer, changing from cobble, to gravel, to very coarse sand-sized material. This transition in grain size is one bit of evidence that the ancient environment of deposition that created this part of our stratigraphic section was rapidly changing. When environments remain the same, so do the rocks deposited in them. Any change in rock type tells us that the environments, which control rock type, must themselves have somehow changed.

In these oldest of Sucia strata, we find only small shards of fossil shells. They are so fragmented and abraded that they can't be assigned to any genus or species. Nevertheless, they can still tell us something about the ancient environments where these coarse, basal strata were deposited. The fact that there are fossils at all in the conglomerates and basal sandstone, and the fact that they are all fragmented, are important clues. They tell us that the basal Sucia strata accumulated in a high-energy environment at the edge of the sea and that these fossils were transported by currents and broken by heavy wave action. This biological evidence is consistent with the clues given by the nature of the sediments themselves, for the gravel and coarse sandstone show unmistakable signs of water movement. The pebbles of the conglomerates are all stacked in a rough order—a process called imbrication that occurs only when small rocks are deposited in moving water. The sandstone too gives evidence of having been deposited in moving water; it shows the sedimentary structures called cross beds, the fossil remains of ancient sand dunes or ripples, both of which can form only in moving water. From this evidence, we thus arrive at a convincing picture of a seashore where strong waves act on gravel and sand. The fossil material was probably brought in from deeper water by this wave action.

As we resume our stroll along the ancient sedimentary beds that now make up the Sucia Island shore—and thus move ahead in time—we begin to see marked changes in the fossil evidence. We are now well into the sandstone beds. The fossils here are still fragmented, but now there are far more of them than below, in the conglomerates, and here in the sandstone they are concentrated in distinct beds. These beds are packed with the remains of small snail and clam shells. The clam fossils look a lot like clams found in today's shallow seas. The layers they sit in look much like sedimentary beds sculpted today during large storms, and we can assume that these ancient beds were formed in similar fashion, during large storms striking the ancient seashore that was Sucia Island, more than 75 million years ago.

As we continue our trek, we observe that the coarse sandstone begins to change subtly in appearance. There are fewer of the shell layers and more fossils occurring singly or in small groups. The fossils appear better preserved as well, and some are much larger than any we have seen before. The majority of fossils are clam shells, but there are many snails as well. Two types of fossils are most noticeable because of their large size and abundance: long prism-shaped clams, some half a foot long, and other clams with a very knobby and sharp-ridged appearance. The first of these is a clam known as *Pinna*, which can be found in abundance today in many tropical regions of the world. The second is *Trigonia*, found today only in Australia. The presence of these two fossils is another type of clue: It alerts us to the possibility that the ancient Sucia environment was far more tropical than now, that these fossil lived in shallow seas of the type we find in the coral reef latitudes. When we examine the snail shells here on Sucia, we arrive at a similar conclusion. Most of them are made up of species that today are found only in shallow, warm, tropical seas.

Now we move faster through time, ever upward through this pile of stacked strata, and the sand gives way to a finer sediment. All vestiges of the shallows begin to disappear: No longer can we see any evidence of storm beds, no longer can we see the ripples that waves might forge. It is as if we are standing at the bottom of an aquarium that is slowly being filled, and the level of the sea rises above our heads.

The fossils change as well. The large number of clam fossils, made up—until now—of many types quite familiar to us, become dominated by larger, flat clams that resemble oysters. These are the inoceramid clams, perhaps the most ubiquitous species of the Cretaceous Period and a form that went completely extinct at the end of the Cretaceous. They are found singly in the Sucia sediment and are often encased in concretions, the hard mud balls so characteristic of deeper-water deposits. Ammonites now appear for the first time—mostly the straight *Baculites*, but other species are found as well. Larger coiled varieties are found, some smooth-shelled, others ornamented with an astonishing array of ribs and knobs. How did these shells get here? Ammonites were swimmers like fish, and they used their astonishing, chambered shells to maintain buoyancy much as a submarine does. They must have competed with fish as active predators. We can imagine the ammonites swimming above fine muddy bottoms—bottoms covered with worms and clams, crustaceans and urchins, bottoms perhaps 100 feet deep or more—and sinking down onto these muddy bottom after their death.

We are now in a very fine sediment, a silty matrix with many concretions. We are also near the end of our walk, for although more rock is present, we would have to dive to see it. The strata on Sucia can be seen to continue underwater from the edge of the shore, but no observations can be made because of the profusion of marine life growing over these underwater deposits. We have crossed a quarter-mile of beach and have traversed more than 600 aggregate feet of sedimentary beds. One message comes through loud and clear: The ancient environment of Sucia Island saw an ever-rising level of the sea, a sea that first lapped on a seashore and ended at a level far above a deep muddy bottom. Did the sea rise, or did the land subside? And did this change happen only here or all over the world at the same time? To answer these questions, we need to step back and examine a larger area than Sucia Island, and we need to look at a longer slice of time as well. The marine rocks on Sucia are but a small bit of the much thicker pile of strata that make up the Nanaimo Group. Let's us look at the lithologies of this great pile of strata in the hope of extracting more information about the ancient environment of this place.

Sea level change and ancient environments

Just across the international border from Sucia Island is an assemblage of is-lands big and small named the Gulf Islands. They are almost entirely made up of the same sedimentary rock that constitutes the northern tier of the San Juan Islands. (The name change is purely a matter of nationalism, not geo-logical history.) The seaway in which these islands now float was carved by the glaciers of the last 2.5 million years, which repeatedly slid down through the regional bedrock to carve out this spectacular fjord-like real estate. But the actual creation of the *rock* that makes up these islands took place much longer ago, near the end of the Age of Dinosaurs, as we have seen.

If we take a boat tour through these islands, we can see the entire pack-age of ancient sedimentary rock that makes up this assemblage, which is named the Nanaimo Group. It is really a group of formations, each formation being defined by a distinctive lithology of rock type. About eleven or twelve formations have been recognized, and their aggregate thickness would be more than 15,000 feet—3 miles of strata—if they were piled up in one place. The name Nanaimo Group comes from a large town on Vancouver Island where these rocks were first identified. The lowest strata of the Nanaimo Group are composed of gravel and sandstone quite coarse in nature, and our best guess is that they are about 85 million years old. There is abundant fossil and litholog-ical evidence that these basal Nanaimo Group strata were deposited in rivers or on land. Perhaps it was a river valley or perhaps a flood plain; whichever, these oldest deposits were certainly not laid down in the sea. But overlying these continental deposits, which themselves sit upon far older volcanic rock, is a dark shale that is indeed the remains of an ancient ocean bottom.

These lowest *marine* rocks are called the Haslam Formation. They are widely exposed over much of Vancouver Island, as well as being found on members of both the Gulf and San Juan Islands. The Haslam Formation is usually more than 1000 feet thick and rich in fossils. The most abundant of these fossils are small forms called inoceramid clams, a type now extinct. But also found in substantial numbers are ammonites. At least twelve species occur, sporting the disc-like shells as well as curious, uncoiled ones. Some look like candy canes, others are shaped like snails. Some are completely un-coiled, others only slightly so. This rich diversity proclaims that ammonites, even this late in their history, were wildly successful creatures. During the time that the Haslam Formation was being deposited across much of the Vancouver Island region, they proliferated.

The presence of ammonites in the Haslam Formation is a sure sign that these rocks were formed in the sea, not on land. The ammonites tell us that the Haslam Formation can be approximately correlated to rocks in the Western Interior that contain ammonites and ash beds. Those ash beds are radiometrically dated at over 80 million years in age. This figure gives us a sense of the maximal age of these sediments on Vancouver Island.

The Haslam Formation can be seen in river cuts and shorelines, and it is usually tilted as a result of regional land movement and mountain building. As on Sucia, when the formation is tilted, we can most easily measure its thickness and most easily observe the rocks that succeed it. Any package of sedimentary rock was originally like a layer cake, and the Haslam Formation is like a deep, dark chocolate layer close to the bottom of our cake. It is overlain by rocks very different in appearance.

Near the top of the Haslam shale a curious thing happens. The rocks begin to change grain size. The fine mudstone and shale gradually change to sandstone which is then abruptly overlain by an enormous thickness of conglomerate that looks somewhat like the conglomerate on Sucia but is usually composed of larger cobbles. The transition is dramatic. For whatever reason, the environment where, for more than 1 million years (and perhaps as long as 5 million years) only fine mud particles accumulated on the bottom of the sea suddenly began to receive gravel, pebbles, and larger hunks of rock. This transition clearly records a dramatic change in the environmental history of the region. It also marks a change in formations. The top of the Haslam Formation is marked by the last marine shale. In many cliff faces and river cuts, we can observe the rocks that succeed the shale; they are composed of very prominent conglomerates. This lithological transition defines the boundary between the formations.

In places, these conglomerates (which are called the Extension Formation), are also as much as a thousand feet in aggregate thickness. Where this formation is exposed, as on the south face of Waldron Island, we see only a huge wall of rounded rocks all jumbled together, with only the faintest hint of sedimentary bedding. Near their top we see them thin and change to smaller clast sizes, eventually to be succeeded by siltstone with fossils.

If we continue our tour through the Nanaimo Group, we find at least five such transitions from fossil-bearing shale to more coarsely grained deposits without fossils. This alternation between marine deposits and marginal marine or terrestrial deposits that makes up the Nanaimo Group continued until the end of the Age of Dinosaurs. The section on Sucia is roughly in the middle of this long history.

Why do such dramatic changes in lithology take place? Normally only two explanations are offered. Either the level of the land is rising or falling (which happens when mountains form nearby), or the sea level itself is changing. It is the latter explanation that is now favored in most cases. If this explanation is correct, the level of the sea rose dramatically over a 20-million-year period and then receded at least five times. Sucia Island was deposited during the third of these cycles. Are these cycles the result of a change in sea level?

Every day the sea rises and falls, its tide a clock of nature. Although we think of tides as coming in and going out (we have even invented two words for the phenomenon, *flood* and *ebb*), the change is more a vertical than a horizontal effect. The tides are familiar to all of us as a normal consequence of gravity, a large moon so near, and a sun tugging on everything. Tidal change is usually imperceptible—unless you live in the Bay of Fundy, where it takes a steady trot to outrun the tide. The tides are the most familiar change in what we may call *sea level*.

Yet daily tidal change is not the only routine change in sea level; it is only the fastest. There are longer-term changes that are far less intuitive and familiar. Off the east coast of North America, for instance, fishing boats routinely trawl up cobbles and rocks indicative of the seashore from depths as deep as 150 meters. The teeth of mastodons and mammoths are found in this way as well, and unless these great Ice Age elephants of North America were swimmers far from shore (doubtful), or had their teeth somehow transported to locales many miles off the shoreline (equally doubtful), we can conclude that during the Ice Ages, the level of the sea was far lower from what it is now.

The sea has indeed been quite unstable in its position relative to the land's surface. Sea level routinely changes, although the rate of change is so slow that no change would be observable in a human lifetime, or even many lifetimes. How does this happen? How did science make this discovery?

That the sea does change in level over time has been known for nearly two centuries. When nineteenth-century geologists discovered the reality of the Ice Ages—when it became clear that the enormous quantity of ice covering so much of continental land surfaces could have come only from the sea—they realized that the consequences of such an event would be a lowering of sea level. So large are the volumes of water required that fresh water on land could not have produced the necessary volumes of ice. Large portions of water were extracted from the sea to be turned into ice. Normal runoff back into the sea did not take place. More and more water became continental ice, and the level of the ocean began to fall.

The most recent Ice Age was not a single event, but the alternation of more than 20 individual glacial advances and retreats covering the last 2 million years of earth history. Each glacial advance and retreat caused a corresponding retreat and advance of sea level. What is so extraordinary is the rapidity and magnitude of these changes: drops and rises of as much as 150 meters.

We have arrived at two causes of change in sea level—tides, which cause changes of up to 15 meters within a single day, and continental glaciation (and growth of icecaps), which causes changes over thousands of years. But there is a third type of sea level change that is far slower. It is currently the subject of intense research by a phalanx of earth scientists, and the impetus for this sustained research comes from the oil companies. According to many earth scientists, this third type of sea level change is brought about by long-term changes in the volume of the ocean basins. These changes, called eustatic sea level changes, take place over time scales measured in hundreds of thousands of years. Eustatic changes appear to be caused by the same processes that bring about continental drift.

Through plate tectonic processes (so goes the most popular theory), the volume of the ocean basins increases or decreases, and these volumetric changes are sufficient to cause the oceans to spill onto land surfaces and flood the lowland areas of all continents. The mechanism that causes this volumetric change is the rise and fall of mid-ocean "spreading centers," such as the mid-ocean ridge extending the length of the Atlantic. We know that sea floor spreading is caused by convection of the earth's mantle region and that the giant plates (which can include both oceans and continents on the same plate) sit atop these gigantic convection cells. Creation of new oceanic crust occurs at the spreading centers, and if the rate at which heat rises in these regions increases, the spreading centers themselves increase in volume. When heat flow diminishes, the spreading centers contract. They act like gigantic mountain ranges that inflate or deflate in response to the heat within them.

The changing volume of rock within the spreading centers causes the changes in sea level. When spreading centers receive more heat and increase in volume, the net effect is to *decrease* the ocean basin volume and cause water to flow out of the oceans onto land surface, much like a bathtub overflowing. Sea level rises. The opposite occurs when heat flow decreases, and the spreading center subsides. Sea level drops.

The magnitude of these changes can be enormous. Near the end of the Age of Dinosaurs, the sea may have been at its "highest" stand ever; half of North America, as well as huge regions of all other continents, were covered with shallow inland seas. North America was really two continents: one east, one west, with a seaway extending from the Rocky Mountains to near the Appalachians and from the Arctic Circle to the tropics. And because the "seven seas" are really a single global ocean of interconnected water, *any* change in sea level is global.

The concept that long-term global changes occur in sea level has launched a rising star among the geosciences; the field of sea level change. A number of new terms have been coined to describe all the manifestations of sea level change: high stands (times of sea water high levels), low stands (the opposite), onlap (a landward migration of underwater deposits as the sea slowly floods a continental surface), offlap (the opposite), transgression (a slow flood of continental regions caused by rising sea level), and regression (the opposite). The field of sea level change remains one of the most intensively studied disciplines of geology, because it has been so useful in finding oil. It is also useful in determining both time and the character of past environments.

A quantum leap in our understanding of sea level change came about in the 1960s and 1970s with the introduction of "seismic stratigraphy." Geologists discovered that by tracing the path of shock waves moving through rock, they could observe structures deeply buried under the earth's surface. When charges of dynamite are exploded above a buried pile of sedimentary rock and the reflections of these shock waves are followed through the rock with recording instruments, very detailed three-dimensional pictures of the stratigraphy emerge. Using this methodology, the geologists recognize onlap and offlap deposits in sediment now deeply buried. The oil companies have found (and still find) a great deal of oil in this way. Over many years, scientists also noticed that in many different parts of the world the sea level curves—diagrams illustrating sea levels through time—seemed to match up.

But do the various sea level curves really match up, or are we seeing different curves in every individual ocean basin? To find out, you need excellent time control correlation matching the various sedimentary rocks that record the sea level changes—in the other words, you need to be able to recognize the age of sediments in widely disparate regions. Without great time control, the acceptance or rejection of global sea level curves could not be completed. By some strange coincidence, the latter part of the Cretaceous seems to hold as much oil as, or even more oil than, rocks of any other age. Yet for reasons outlined in the first several chapters of this book, the sea level curves for the latter part of the Cretaceous Period have been among the least well known, because it has been so difficult to find and decipher the Cretaceous time markers on a global scale.

Five sea level oscillations are present in the sedimentary rocks of Late Cretaceous age in the Vancouver Island region. Are these regional or global changes? Only by studying rocks of similar age in many other regions could this be ascertained.

Sea level change and life

Sea level change is certainly an important environmental factor that affected ancient environments, but it may play an even more important role in governing the nature of life on earth. For more than a century, sea level change has been considered one of the leading causes of mass extinctions, which turn out to be the largest-scale evolutionary phenomena that affect our planet's biota.

Mass extinctions are global catastrophes that have caused large numbers of species to become extinct in the geological past. A mass extinction event can last between a thousand and several million years. There have been about 15 such events during the last 500 million years of earth history (the time of skeletonized life on this planet). Five of these are classified as "major" in that they caused more than half of all species then living to go extinct. The most destructive mass extinction ended the Paleozoic Era. The second most destructive ended the Mesozoic Era, of which the Cretaceous is the last unit. This "K/T" mass extinction ended the Age of Dinosaurs and brought the Cretaceous Period to an end as well.

The possible association between sea level change and mass extinction was recognized early by geologists. More than a century ago, it was noted that each of the major mass extinctions seemed to have occurred soon after a sudden and precipitous drop in global sea level. Yet by what mechanism might a simple drop in sea level have caused planetary death on a global scale? There is no doubt that several of the major mass extinctions occurred either during or immediately after a great sea level change. Might this have been only simple coincidence?

The rising or falling of the sea's surface would probably not itself kill anything (unless, of course, it caused shallow seas to disappear, thus killing all of the trapped inhabitants). However, the change in sea level could trigger a much more deadly killer: catastrophic global climate change. As sea level drops, land area increases relative to the area of the sea; this, in turn, stimulates climate change, because temperature fluctuations on the earth's surface are significantly influenced by the relative amounts of ocean and land area. Sea level change can also reconfigure the geography of the earth, and this too plays a part in the nature of oceanic current systems. Because organisms are usually quite narrowly adapted to specific climates, sudden climate change is a plausible "kill mechanism" associated with sea level change.

Geologists of my generation were thus faced with several very interesting questions about sea level change. Was it global? And did a major sea level change occur at the very end of the Cretaceous Period and thus create a climate change that exterminated the dinosaurs and much else? These are two different questions, and neither could be tested by using observations of rocks only in one area. Furthermore, the fact that the Cretaceous rocks in the Vancouver Island region occurred on either isolated islands or isolated riverbeds left much of the complete record of time in this region covered by the sea or by a sea of vegetation. A different succession of sedimentary rocks of this age needed to be studied, rocks with better complete exposure, a place where neither the sea nor vegetative cover was present. A desert region was needed, but a desert region near the European type regions so as to ensure good time correlation.

After some deliberation, an international group of geologists chose a thick exposure of sedimentary rocks found in central Tunisia. It was decided that an expedition would be mounted to look at these rocks in order to determine whether the mass extinction event at the end of the Cretaceous was coincident with a major drop in sea level. I was invited, and I was eager to go for many reasons, not the least being to see whether the sea level changes I had long observed in the Vancouver Island area could be recognized in Tunisia as well.

The verdict

All the others had gone, leaving me alone in the empty hotel in El Kef. It was very cold.

I had come to El Kef, Tunisia, in April of 1992 to help conduct the crucial observations concerning sea level change and the "K/T" mass extinction represented by the region's strata and fossils. The rocky country did not disappoint. Tunisia has some of the best exposures of the Cretaceous/Tertiary boundary known in the world, so it had been chosen as the site of a test to determine whether the great mass extinction that ended the Age of Dinosaurs—thought to have been caused by climate change, or sea level change, or even by the collision of a great comet with the earth—was a sudden, short-term or a slow, more gradual event. Two dozen scientists had come here to collect sedimentary rock from either side of the Cretaceous/Tertiary boundary in order to see whether the enclosed fossils disappeared right at the "K/T" boundary or did so gradually well below the boundary and to see whether the geological evidence suggested that the extinction coincided with a drop in sea level, as so many had postulated. If the latter were true, then a sudden drop in sea level just might be one of the most extraordinary killers on the planet.

We were especially interested in the nature of strata found in the last meters of the Cretaceous. If sea level were dropping here, we would expect to find evidence of shallowing water, and thus coarser-grained strata, as we approached the "K/T" boundary and overlying Cenozoic strata. Our methodology, our "time machine," was the same as that used on Sucia Island in the account earlier in this chapter. We "walked" the section, making detailed measurements of the stratal thickness and minute observations about every aspect of the rocks: their grain size, their color, the types of fossils present (or absent), the nature of sedimentary structures such as ripple marks or cross beds, the presence or absence of conglomerates, and any trends discernible in these features. No high technology was needed—just detailed observation carefully recorded in field notebooks.

After several days of this our results were clear. There was no evidence of a sudden drop in sea level coincident with the great mass extinction. In fact, we found just the opposite. The water in this region (and in many others, it turned out) was *deepening* when the extinction occurred, not shallowing. Sea level was not the culprit. Abundant evidence eventually showed that earth's collision with a great comet in the Yucatan region of Mexico was the real killer.

At the end of a week the various geologists left for their home countries. I stayed on to spend more time in the field, for my work here was still unfinished. The hills in this part of Tunisia contain thick successions of sedimentary rock exactly the same age as the sedimentary rocks in the Vancouver Island region. Here was an ideal test of the hypothesis that sea level change is nearly always caused by global events rather than due solely to local events such as regional uplift, which might be caused by mountain building. Here I could see whether the pattern of sea level change so wonderfully exposed on the various islands of the Vancouver Island region was caused by global rather than regional processes.

El Kef is located in the central part of Tunisia and is at sufficient altitude that spring comes late. It seemed ironic to be so cold here, for to the south of us lay broad desert regions, giving way to the parched and hot Sahara. It was a small town, with few of the Western trappings that characterize the larger cities farther north in Tunisia. A truck with a loudspeaker drove through the town each morning and evening, calling the faithful to the mosque. Arab men thronged the bars and cafes, but few women were to be seen. I could feel the discipline of the place, and its strangeness to me was palpable.

The days in the field were long and exhilarating. The rocks were laid out as if by giant brushstrokes across the face of the low mountains, alternations of thick limestone and dark shale streaked across the rugged countryside. Ammonites had been reported from here but had barely been studied. I had assumed they would be rare and so was delighted to find them in abundance. In places they were almost stacked together and elsewhere they were rare, with no apparent reason for their curious distributions. The limestone contained the most ammonites, and it was there that I spent most of my time, scanning rock surfaces for the telltale sign of ribs or a spiraled shape. The most common forms had shells that looked like giant snails, rather than the usual coiling in one pale that typified the vast majority of ammonites. One particular species was especially common, a form called *Bostrychoceras polyplocum* well known from Europe. Rocks containing this particular ammonite may be exactly time-correlated with the rocks of Sucia Island, and like the rocks on Sucia, the sediments encasing this particular ammonite showed evidence of increasing water depth. Limestone and shale: a dance of ancient rock, one telling of a deepening sea, the other of dropping sea level.

The Tunisian rocks were beautifully exposed; there were no missing intervals due to faulting, no covered intervals due to vegetation. In this regard they were completely unlike the rocks I had long studied in the Vancouver Island region, where the strata are found only on seacoasts or in river gorges. Here, in this Tunisian semi-desert, they are inescapable. They are the land itself, the skeleton of this country with the bones exposed.

The very abundance and completeness of the rock record here enabled me actually to *see* how important sea level change is in determining the nature of sedimentary rocks. There appeared to be many different cycles of change superimposed on one another. Long-term changes visible in thousand-foot increments of sediment (and thus talking place over millions of years) were made up of shorter-term fluctuations visible in hundred-foot increments. These changes could be read in the way the rocks looked, in the way they were bedded, and in the way their enclosed fossils changed. I could see shallow-water organisms being replaced by deeper-water assemblages as I moved up through thousands of feet of strata in this region, and then I could watch the opposite occur. The level of the sea had been the master here, dictating so many aspects of the formation of sedimentary rocks long ago. In turn, the ancient rocks themselves dictate how the land looks today, not just in the colors of the dark shale and white limestone, but the very nature of the landscape itself. The limestone is far harder than the shale and less easily eroded; because of this, the limestone makes up the ridges and high places, and the softer shale is usually found at lower elevations. You can literally map the topography by mapping the limestone and shale, its locations brought about by ancient changes in ancient sea level.

Worldwide units. It seemed the very essence of circularity. It took high-precision, worldwide correlation through fossils and radiometric dating to convince scientists that the fluctuations in sea level, as deduced from the rocks left behind, were worldwide in scope. Yet once that had been proved, the very changes in sea level—the worldwide sea level curve—became a new sort of clock, a new way of telling time. Once you find your place in the sea level curve, you can make powerful predictions about time.

It would seem that there could be no two places more dissimilar than the green islands of the Vancouver Island region of North America and the deserts of northern Africa. The rocks are equally dissimilar: the dark olive sandstone and shale of the Nanaimo Group and the white limestone of Tunisia. Even the fossils in both regions are different, and no species common to both exists. But they are united by their age: Both sets of rocks were deposited between 80 and 65 million years ago, and being of the same age, they experienced the same fluctuations in sea level. The simplest of all time machines, detailed observation, has shown us that this is so and has elucidated two ancient places and their environments.

Part Three

Inhabitants

6

The Bite of a Mosasaur

Far up the Inside Passage, Vancouver Island begins to nestle inward toward mainland British Columbia, and the frigid water begins to warm a bit as the Georgia Strait narrows. The salmon fishing improves, and the feel of wildness increases. Many small towns dot the countryside, most surviving on logging, tourism, and the groceries sold to folks in retirement double-wides. It is certainly not the U.S.A., for the subtle but distinct feel of Canada is perceptible everywhere.

The nature of the rocks begins to change as well. Fossils in this part of Vancouver Island are no longer found on the seacoast. To find fossils this far north, you have to trek inland on the logging roads winding through the second- or third-growth forest, where isolated canyons of shale have been cut deep into the countryside by the region's many rivers and creeks. The names of these watercourses reflect a heritage of native Canadians and logging moguls: Tsable, Bloedel, Trent, Puntledge, Browns, Nanaimo, Qualicum, Little Qualicum, Haslam, Cowichan, Chemainus; each is a river where the dark shale and their enclosed ammonites lie exposed. All have rocks older than either Sucia or Hornby, and the ammonites are different, too. The two most common are uncoiled species—one a giant snail shape, the other an open spiral; both are exquisite to find and behold, real heart thumpers. Elsewhere in the dark strata are many types of clams and snails, but in forms associated with deep-water environments.

I came with a crowd in 1994 to see these creeks and their fossil fauna: friends from the local Seattle museum, my son, and a Seattle reporter named Roger Downey. All had come for a week of fossil collecting. We camped along the rivers and excavated in the dark shale, moving northward up the east coast of Vancouver Island and its Nanaimo Group deposits. It was a perfect lazy summer trip; no pressing agenda, beautiful fossils, a good group. One of our last stops was the town of Courtney, sprawled along the banks of the Puntledge River. Courtney sported something none of the other small logging towns had: a museum. And in this museum reposed the most extraordinary fossil ever collected in British Columbia: a 50-foot-long elasmosaur.

Although the Age of Dinosaurs is renowned for its saurian behemoths on land, the seas supported large and exotic reptilian wildlife as well. Great sea lizards and long-necked reptiles occupied roughly the niche of today's toothed whales. Yet prior to the late 1980s, not a single bone of any of these creatures had ever been found in the Vancouver Island region, and it was thought that perhaps these great sea serpents had not lived in this region of the world, 80 to 70 million years ago. That perception was radically changed by an amateur fossil hunter prowling the banks of the Puntledge River.

The skeleton was extraordinary, as was the story of its recovery. A local resident, Mike Trask, was wandering the edge of the Puntledge, looking for ammonites, when he spotted a most unusual fossil just sticking out of the riverbank. It turned out to be the vertebra of a large marine reptile. Trask soon spotted other vertebras, and various other bones as well. Large excavating equipment was brought in, and over several months the remains of this creature—intact and complete—were removed from its lithic tomb. Specialists from the Royal Tyrell Museum in Alberta were brought in to help with skull preparation, but it was largely a local production. The best part of this story is that the skeleton stayed home. It was not sold for big bucks; it was not exiled to some far off museum. It stayed in Courtney, and a museum grew up around it.

This particular find galvanized the paleontological community in British Columbia and elsewhere. Simply knowing that such fossils were present in the Vancouver Island region sharpened eyes grown complacent and brought new collectors to the region to search for this type of fossil treasure. Before long several more finds were made. Perhaps the most exciting came from Hornby Island, at the same site where we had drilled for paleomagnetics. On Fossil Beach, one of the best of northwest collectors, a man named Graham Beard, found the region's first mosasaur fossil.

The presence of a mosasaur, a large marine lizard, sent anyone who had discovered ammonites in this region scurrying back to look at their fossils, for mosasaurs are thought to have been the most important of all ammonite predators. This conclusion had come from supposedly telltale bite marks found on many ammonite shells.

The past *is* time and place. But it is life, the inhabitants, as well. The fossil themselves that we find are time machines, for they as much as any other information can transport us back into deep time. But as we shall see in this chapter, the study of fossils is fraught with uncertainty and with multiple paths of interpretation. The fossils are the data: They are the real objects coming to us from deep time, and as is written on many a laboratory door, "Good data are immortal." But data are as naked as the skeletons we find; they need to be prepared and interpreted. Though the data are immortal, the interpretations are not. To put interpretations to the test, scientists all over the world invoke another time machine.

Evidence of ancient predation

A seashell pressed to the ear will evoke the sound of surf, aural waves on some far tropical beach, the tone of the sea. But any seashell records far more than an imaginary surf. Shells bear the equivalent of a tree's rings, giving us age and clues to environment; all shells maintain a record of their life and times, and sometimes the time and manner of their death. One of the properties of sea shells—including *ancient* sea shells—is that they are faithful recorders of age-old predation.

Much information about the past comes from reading the record of shell *breaks* that offer an account of some ancient predatory attack. Interpreting breaks in fossil shells is an important and widely used means of bringing the past back to life. The shells and their breakage patterns, then, are the data. But our interpretation of what caused those breaks can be far more ambiguous.

Many fossil shells show breaks. Sometimes the attacks that caused these breaks killed the animal within. Often, however, the predator did not succeed in killing its shelled quarry, leaving the organism to repair the breaks and, in the process, leaving a scar behind, yielding fascinating information about who ate whom in ancient time. Sometimes, however, the investigator's zeal for such investigations outstrips the evidence. And when such excitement is coupled with an incomplete or faulty understanding of how shell material fails under pressure or point load, misinterpretations can ensue. The following is just such a cautionary tale. It not only pitted those who understand the mechanical and material properties of molluscan shell against those who don't but also cast into sharp relief a clearly sensational explanation for fossil evidence against a far more prosaic—yet in this case accurate—portrayal. Ammonite shells often have round marks in them. Were these holes left by the dramatic attack of a marine lizard of great size and strength or by the slow rasping of a tiny snail shell over many years? Are these holes the bite marks of a creature whose closest living relative is the Komodo dragon?

The Komodo dragon is a large lizard (up to 10 feet long) that now lives only on several small Indonesian islands, where it preys mainly on small deer and pigs, killing them with a saliva rendered poisonous by highly toxic bacteria. The Komodo dragon is thus a most impressive creature. However, the modern lizards are dwarfed by their ancestors from the Age of Dinosaurs, which were 30 to 40 feet in length and were, during the heyday of the dinosaurs, the most successful and largest of all seagoing vertebrates.

Although relatively few people have seen a Komodo dragon (and no one has ever seen a living mosasaur, because the last of them died out at the end of the Cretaceous Period, killed off no doubt by the lingering effects of the comet that ended the Mesozoic Era), their family is fairly well known to us. All, the living and the dead alike, are monitor lizards, a group still highly successful in many parts of the world. Long ago, during the Cretaceous Period, giant monitor lizards—the mosasaurs—were the largest and most ferocious carnivores in the sea.

Mosasaurs were not the only large reptiles in the sea, of course. Two other oceangoing reptiles also vie for our attention in the Mesozoic rock record: the long-necked (and sometimes short-necked as well) creatures known as plesiosaurs (the archetype for the Loch Ness monster) and the more fish-like (or dolphin-like) forms called ichthyosaurs. The latter were extremely abundant in the Jurassic, but by the Cretaceous Period they were eclipsed in number (and certainly in size) by the mosasaurs and plesiosaurs (elasmosaurs, such as the type found on the Puntledge River, are a type of plesiosaur). Ichthyosaurs and plesiosaurs were not dinosaurs either—it appears that no true dinosaurs made the evolutionary leap back into the sea.

Mosasaurs were very late arrivals to the Mesozoic world. Unlike plesiosaurs and ichthyosaurs, which can be traced well back into the earliest epochs of the Mesozoic Era, mosasaurs first appeared in the Late Cretaceous. Until the discoveries on Vancouver and Hornby Islands, they were largely absent from the west coast of North America, having been found only at a few localities in California.

All mosasaurs had four flippers, with the toes expanded into web-feet. The bodies were long, the tail even longer. Some had tails with a large caudal fin as well, a structure unknown in our world's lizards. As in modern-day whales, the bones of the pelvic region were reduced, making locomotion on land almost impossible. Yet unlike whales, which have lost any semblance of a neck, the mosasaurs retained distinct head regions set apart from the body by a strong, if short, neck. The head itself was out of some nightmare. Flat and narrow with a long, tapering snout, its most distinguishing characteristic was the numerous teeth. The jaws, like those of most vertebrate animals, were lined above and below with numerous dagger-like teeth. But the palate bones had teeth as well, a second set also conical and tapering well back in the roof of the mouth. Clearly, these were creatures evolved to bite. Finally, there is the matter of the tongue. All present-day monitor lizards have a forked tongue, like a snake's. In all probability the mosasaurs of the ancient world were similarly endowed. No wonder these creatures are favorites of illustrations depicting the Mesozoic seaways, for here is a monster every bit as frightful as any *T. rex* or any of the vermin emerging from the imagination of H. P. Lovecraft. Imagine some sort of odious (but spectacular) gladiatorial combat between mosasaurs and killer whales of our world or between a mosasaur and a great white shark. I would bet on the mosasaurs.

Why did the lizard-ancestors of the mosasaurs go back into the sea? Why abandon the rich, tropical rainforests of the Mesozoic Era, a time when

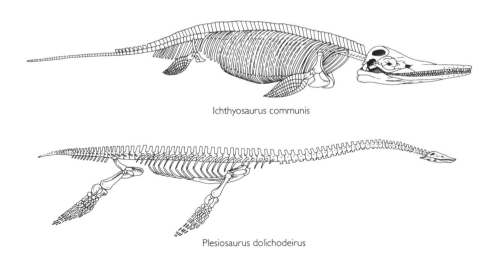

Ichthyosaurus communis

Plesiosaurus dolichodeirus

Fossil Ichthyosaur (*above*) and Plesiosaur (*below*) of Jurassic Age.

jungle largely covered the land, for life in the sea? We might also ask the marine iguanas of our day, who have taken the same step. (Though not so giant a step: Marine iguanas are fully amphibious, being as adept on land as in the sea, whereas the mosasaurs may have left the sea only to lay eggs—or not at all, giving live birth in the sea just like ichthyosaurs and whales.) The answer to all these questions may be the same: food. The sea is rich in resources, and the lure of that food has caused many lineages of animals to return to the place of their ultimate ancestry. And if it was food that induced a previously obscure lineage of small monitor lizards to abandon the land in the Late Cretaceous, what was the nature of that food? That particular answer is on display in virtually every natural history museum on earth. The seas of the Mesozoic were filled with those ancient shelled cephalopods called ammonites. Uncounted numbers of ammonite shells have been recovered from Cretaceous-aged deposits bearing numerous circular holes that closely match the size and shape of mosasaur teeth. Case closed: Mosasaurs ate ammonites and left a record of their feasts behind, dramatic statements of ancient predation.

And so it is that countless natural history museums, if not lucky enough to have a mosasaur skeleton to display, at least can show an ammonite shell bitten, supposedly, by a mosasaur. The only problem is that in every case examined to date by a team of Japanese and Canadian scientists that has scrutinized this famous paleontological lore, the so-called mosasaur bite marks are anything but the work of gnashing mosasaur teeth. In fact, they have a

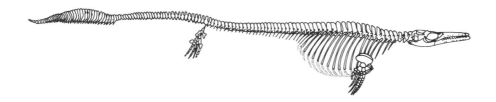

Fossil Mosasaur of Cretaceous Age.

far more prosaic cause: They appear to be nothing more exotic than the resting holes of circular marine snails called limpets; resting holes are small depressions that limpets drill into the shell or rock that serves as a home territory. Such a radical reinterpretation has predictably caused howls of protest. What is the origin of this interpretation, first made in 1960, that the circular

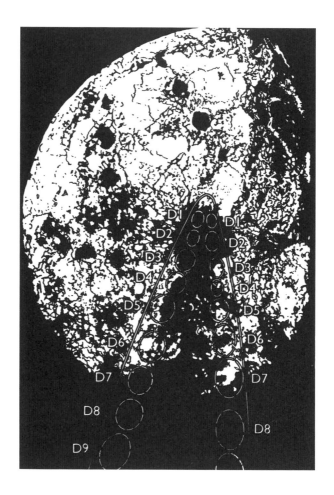

Diagram showing ammonite shell and position of so-called mosasaur bite marks. (From Kauffman and Kesling, 1960.)

holes found on so many Upper Cretaceous ammonites are mosasaur bite marks, and why (or how) was it shown to be wrong?

A tempting interpretation

The story began in the early 1960s and was originated by one of the most flamboyant and colorful characters ever to have studied the deep past, Erle Kauffman. Erle, as he is known to all, remains one of the most revered of living paleontologists. He grew up in the west, and during his long career he became perhaps the finest field paleontologist of the twentieth century, specializing in the paleontology and stratigraphy of Cretaceous rocks. His major study area has been the great Western Interior Seaway of the United States, the large sea that split the land area of North America in half, running from north to south. This seaway was the home to uncounted ammonites, and not a few mosasaurs as well, judging from the large numbers of skeletons that have been collected over the years from formations such as the Pierre Shale and the Niobrara chalk. Somewhere, early in his career, Erle came to the conclusion that mosasaurs must have eaten ammonites.

The logic is impeccable. The very abundance of ammonites in Cretaceous oceans must have made them a tempting and attractive food source for many types of marine carnivores. Because many ammonites were large (many species had shells over 3 feet in diameter), they would have been a very appropriate prey for the mosasaurs—providing, of course, that the hard outer shell of the ammonite could be breached.

Erle and his academic advisor A. Kesling at the University of Michigan first published this interpretation in 1960. The paper remains sprightly and readable. It starts as follows:

> An Upper Cretaceous ammonite of the genus *Placenticeras* has been found bearing numerous perforations and impressions made by the teeth of a mosasaur. . . . The shell was bitten repeatedly, and bears dramatic evidence of the fatal encounter. From a composite

135

of the patterns of tooth marks we have reconstructed the dentition, and from the relationships of the upper and lower jaw we offer certain inferences on the shape of the head, the structure and action of the jaws, and the diet of the mosasaur. (Kauffman and Kesling, 1960, p. 193)

This scientific paper about the proposed mosasaur bite marks found in ammonite shells was so different in its style and tenor from the paleontologic papers of its time that it stands out like a beacon. Paleontology in the 1950s had been dominated by descriptions of ancient species, not descriptions of how those species lived, but here was a true interdisciplinary paper. Not only did Kauffman and Kesling dare to describe mosasaurs (the province and territory of "vertebrate" paleontologists), but in the same paper they described ammonites, an "invertebrate" fossil. Specialists in the two fields at that time (and to some extent even now) rarely read each other's papers and don't even attend the same scientific meetings. Was this a paper about mosasaurs, or ammonites, or what?

The paper was also fun to read. Some of the descriptions sound far more like a novel than like a learned scientific paper. Mosasaurs are characterized as having a "pugnacious nature" and are described as "the most vicious, rapacious creatures of the warm epicontinental seas during the epoch of their existence."

The Kauffman and Kesling paper dealt with *one single fossil*—and still managed to come in at a length of 58 pages and to offer 7 figures and 9 full-page plates of photos. Perhaps never in the history of paleontology has so much been written about a single fossil shell. The attack itself is described in blow-by-blow fashion, for Kauffman and Kesling believed that their shell (which is indeed riddled with holes) was bitten sixteen times. Each bite is described in detail. For instance,

The initial bite resulted in eight impressions on the shell. Teeth of the upper jaw struck the right side of the conch, leaving marks of maxillary teeth. . . . According to our reconstruction of the mosasaur's dentition, based on a composite of all bites, all teeth represented in bite 1 are strongly developed. Since only three perforated the shell, the mosasaur did not exert much pressure in this bite. . . . Since the bite was directed for the anterodorsal side of the ammonite shell, above its aperture, the animal may have been picked up from the bottom or attacked while swimming at some distance below the surface.

And so on, for fifteen more bites.

The authors concluded that the attack was directed at the upper side of the ammonite. They thought that the mosasaur tried to swallow the entire ammonite, pulling it as far back into the throat as possible, and that the mosasaur was in the habit of eating ammonites.

The plan of assault—seizure from above, attempt to swallow whole, and failing this, crushing the living chamber to squeeze out the soft parts—seems to indicate familiarity with the prey. The large, conical biting teeth and the wide intermaxillary angle appear to be adaptations for catching, killing, and swallowing or crushing big shelled animals, such as ammonites. (Kauffman and Kesling, 1960, pp. 234–235)

All in all, the paper was a sensation, and it launched Erle's career. There was only one huge, glaring problem: Was this the only mosasaur in the history of the world that ate ammonites? Kauffman and Kesling conceded that this was the only specimen known with bite marks. Surely, if one mosasaur was such an avid (if incompetent) ammonite eater, then other shells bearing other telltale bite marks should be known.

Thirty years after his initial publication on this subject, Erle Kauffman was still championing the idea that mosasaurs bit ammonites. By this time he had identified another 30 specimens with "definite predation marks" and at least this many with "suspect" marks. Furthermore, Kauffman asserted that most of the confirmed predated ammonites were attacked in a manner similar to that described by Kauffman and Kesling, but "with greater efficiency," and he also noted that

> In most cases the initial crippling bite is directly followed by a major bite or bites across the back of the ammonite's living chamber, and rapid extraction of the soft parts from the shell. . . . This seems to show a familiarity with the prey and a learned predation pattern; this and the increasing number of known predated ammonites in the Late Cretaceous, suggest that mosasaur predation on ammonites was a normal but not dominant food chain relationship. (Kauffman, 1990, p. 185.)

Mosasaurs and me

Two events in my life had sensitized me to the issue of mosasaurs and ammonites. The earlier had occurred more than two decades before, during my first research trip to New Caledonia. My companion on this trip, a zoologist named Arthur Martin, placed a living nautilus in a tank full of large marine turtles. Nautiluses are not ammonites; the nautilus shell is thicker, and the septa within the shell are far less complex and convoluted. And sea turtles, of course, are unlike mosasaurs in many ways, not the least of which is their lack of a large mouth filled with teeth. Nevertheless, the result of this attack was quite impressive and was fatal to the nautilus, whose shell was fragmented into pieces large and small by the turtle's hard beak. The key observation was that the turtle's attack on the nautilus resulted in a fragmentation of the shell—not a hole or gouge left into the shell by the turtle's beak. Hit a porcelain plate with a hammer and you get a similar result, and for the same reasons. Mollusk shell (like porcelain) is composed of limy material that is prone to shattering upon impact.

The second chance event that piqued my interest in the "mosasaur bites ammonite" story occurred near Calgary, Canada, in early March of 1996, when I had a chance to visit the Royal Tyrell Museum. Sitting on the edge of the great Alberta prairie, this modern museum houses the finest collection of dinosaur specimens in the world. But it holds other fossils as well, including a huge collection of ammonites. By far the most abundant of these belong to the genus *Placenticeras*, a large, discoid ammonite from the Upper Cretaceous and, coincidentally, the same genus first examined by Erle Kauffman in his pioneering mosasaur bite study.

Placenticeras is an extraordinary fossil from an extraordinary creature. The fossil shells of this species are large for an ammonite; specimens a meter in diameter are quite common. (The largest ammonite ever collected is about 3 meters in diameter). In certain Canadian deposits they are very common, surely mirroring their abundance in the shallow Upper Cretaceous oceans that covered this part of Canada from 80 to 70 million years ago. The shells are compressed, like giant discs, and often are found with the mother-of-pearl shell material still beautifully iridescent. It is within the shells of this taxon that the great preponderance of circular holes have been found—the same holes originally described by Kauffman and Kesling as mosasaur bite marks. But in the Tyrell there are so many of these shells, collected over the years, that an efficient numbers game can be played: How many of the shells have holes, and do all or any of the hole marks suggest the spacing that a mosasaur jaw would produce?

My guide to the "mosasaur bites ammonite" story, as least as it was being played out in 1996, was Dr. Paul Johnson of the Tyrell Museum, a specialist in Upper Cretaceous bivalves. Some months earlier, Johnson had been contacted by Dr. Tatsuro Kase, a researcher in Japan, asking whether the Tyrell had ammonites with circular holes in them. Kase had been studying Upper Cretaceous ammonites from Japan and had begun to suspect that many of the circular holes found in these ammonites were the innocuous product of boring (literally and figuratively) limpets of the time, rather than the exciting evidence of ancient predatory behavior.

Kase visited the Tyrell's collections, and in every case the circular holes seemed to conform to the sizes and shapes produced by limpets. When I heard this story, I had to laugh—I certainly preferred that the many holes give us a glimpse into predatory sea monsters instead of snails. The clincher, at least to me, was that the patterns of these holes seemed random; only once in a while in this large collection would a shell show holes arranged in a row, looking for all the world like it had been produced by a row of biting teeth arranged along a linear jaw.

A month after my visit to the Tyrell Museum, the issue of mosasaur bite marks in ammonite shells was center stage at a large meeting celebrating the Age of Dinosaurs. The meeting, named Dinofest 2, was held in Tempe, Arizona. It was the second in a now-biannual event exhibiting and discussing dinosaurs. The dinofests last a month and include a gymnasium-sized public display of many fine dinosaur skeletons and other fossils. A three-day scientific symposium held at the end of the Tempe dinofest was attended by the brightest lights in the dinosaur hunters' firmament: the A list, headed by Jack Horner and Bob Bakker, and the B list, which was virtually everybody else still living who ever studied dinosaurs or their worlds. I was included in the latter category, not because I study dinosaurs, but because I study how they died. Naturally, being a specialist on why everyone's beloved beasts are no longer on earth makes one less than popular. No wonder they scheduled my presentation at the end of the last session. Nevertheless, I was happy to have been invited at all, and I attended many of the excellent talks. One I especially wanted to catch was Erle Kauffman's; its subject was an update on how mosasaurs consumed ammonites in those ancient Cretaceous oceans.

The talks were held in a large room packed with professionals and the general public. Erle strode to the podium at the appointed time and began speaking. Erle showed, on the basis of his decades of bite analyses (nearly 40 years had passed since his original paper with Kesling), how mosasaurs would stealthily creep up on the ammonites and then (depending on the particular species attacking or being attacked), dive down, or swoop from behind, or pirouette in some mad dive-bomber attack, whooshing through the water, snatching their prey unawares, giving them a "crack" and darned if they weren't suddenly calamari. In fact, Erle estimated that at least a hundred specimens had been shown to carry undoubted mosasaur tooth marks. Erle delivered a breathtaking performance, and the rest of the audience clearly felt the same way. It was great: diving mosasaurs, dead ammonites, predation in the deep dark Cretaceous seaways. At the end of the talk, Erle basked in the glow of the respectful questions. Bob Bakker chimed in from the back of the room (as he is wont to do), adding his *imprimatur* to the mosasaur bite mark theory. Bakker described how the peculiarly articulated mosasaur jaws allowed these great marine lizards to envelop the ammonite shells before piercing them with that mouthful of cruel, dagger-like teeth. Other workers added their own anecdotes and stories about bitten ammonites they had seen. This was paleontology at its finest, no doubt: serious research about an inherently interesting topic. Unfortunately, although the topic was good, the science wasn't.

Unknown to me at the time, in Seattle, Sophie Daniel, the eight-year-old daughter of a zoologist named Tom Daniel, was being shown fossil exhibits at the local University of Washington museum. One of the exhibits they viewed together consisted of a mosasaur skeleton encased with an ammonite bearing purported mosasaur "tooth marks." This particular ammonite was displayed in a case at about the eye level of Tom's daughter. She turned to her father (who recounted the whole story to me later) and said, "How could there be tooth marks? The shell would break in pieces and not leave tooth marks." Her father looked at the specimen closely and had to agree with his daughter.

In the question period following Erle's talk at Dinofest, I tried to make a similar point. In the past I had conducted simulated predatory attacks on nautilus shells, using pliers (to simulate the predator), and in every case the shell simply shattered. Imagine, I asked the group, trying to drive a nail through a nautilus shell (or an abalone or clam shell, for that matter). Producing a large, circular hole in a mollusk shell by impact of a large, tooth-shaped object is impossible.

At this point the assembled audience (perhaps about 200 people) became somewhat confused and restless. Who *was* this character from Seattle, not even a dinosaur paleontologist, doubting our hero Bob Bakker as well as the mosasaur's ability to bite any *number* of tooth marks into ammonite shells? Heresy! An uneasy rustling filled the room, and I sat down in a hurry.

141

Luckily for the honor of the Mesozoic, Bob Bakker then saved the day (if only temporarily) for the mosasaurs. He opined that he had seen *many* such holes in ammonites, *obviously* produced by mosasaurs, and that although he respected my observations, he had an explanation that might clear things up. Bob began his lecture from the back of the room: "The mosasaur jaw has long been known to have had a very peculiar articulation. They could unhinge their jaws somewhat, in a manner analogous to most snakes, and position their jaws around the ammonite shell. The circular holes were punched through the ammonites shells, cookie cutter fashion, because the mosasaurs put pressure of the shell *extremely slowly*." With this red herring now fouling the water, order was restored by Neil Larson, a professional fossil collector who has probably found more ammonites than any other person in the world. Neil has found many ammonites with the circular holes within them and thus knows whereof he speaks, at least on any subject related to ammonites. He proposed that he, Erle, and Bob Bakker discuss the various specimens they had seen and report to the audience later in the day. This pleased the group.

As good as his word, later in the day Larson stood up to announce the results of this conference. After consulting with his colleagues, he could say that the mosasaur tooth marks in ammonite shells were real. However, they could be confirmed *in only three specimens in the world*, rather than in the hundred or more claimed earlier by Erle Kauffman. And that is where matters left off as far as Dinofest went. But the affair certainly did not end there.

Putting an interpretation to the test

Upon my return to Seattle, Tom Daniel told me of his daughter's discovery, and I told him of the coincidental dustup at Dinofest. Tom laughed, especially when I described Bob Bakker's theory that a slow bite might produce holes where a rapid bite would not. Pressure is pressure, he declaimed, and there is no way a material such as mollusk shell could ever be bitten so as to produce perfectly circular holes. The shell would crack, just as his daughter intuitively divined upon first seeing an ammonite shell with holes in it.

Tom Daniel had a great advantage in this controversy, for he is a specialist in a field known as biomechanics. Biomechanicists are really engineers dressed in zoologist or paleontologist clothing, for it is engineering principles that they bring to the work bench. This clan has done much work on the nature of molluscan shell and on how it responds to pressure. Typical molluscan shell is composed of several distinct layers and is far more than a simple crystal substance. Tiny crystal of calcium carbonate are interlayered with tough organic material of the same composition as fingernail. The two together give far more strength than either one alone. Combined, they form what is known as a *composite material*. Concrete is another type of composite material; it is formed of lime, sand, and rock, which, when combined, form a far stronger substance than any of these materials alone. If you add in reinforcing steel rods to the mix, you approach (in design) a molluscan shell. One great advantage of molluscan shell over concrete is that it is capable of at least limited bending. Yet in many ways, it also acts as porcelain china does: It reaches a certain critical point and then shatters. Large cracks run through the material at high speed, usually producing irregularly shaped surfaces. Punching holes in a mollusk's shell or a china plate can be done only under very bizarre circumstances. Mosasaur tooth pressure does not seem to be one of them. It seems very unlikely that mosasaurs punched holes in ammonite shells.

Why dwell on this story of what is now nearly four decades of interpretation about ammonites with holes in their shells as being caused by mosasaurs? Because paleontology has labored for so long under "stories" rather than science. Erle Kauffman long ago found a shell with holes in it. He deduced that the holes were made by a marine lizard and then, having come to this conclusion, spun out a blow-by-blow tale to support it. Not a single experiment was ever performed. Tom Daniel, on the other hand, after having been alerted to this interesting problem by his young daughter, set out to see whether various hypotheses would stand up to scrutiny. In doing so he employed the scientific method, perhaps the most powerful of all time machines. Tom convinced a grad student named Erica Roux to try actually to *produce* round holes in nautilus shells (very close approximations of ammonite shells) experimentally. This experiment was designed to determine whether the hypothesis that a tooth could break a circular hole in a shell could be falsified (shown to be false).

Erica constructed an artificial mosasaur jaw. It did not look much like the real thing, being fabricated of metal with series of teeth made out of nails and screws, but nevertheless it closely approximated the real thing in many ways. The "teeth" descended onto the shell surface just as a mosasaur jaw would have, and a gauge attached to the jaw showed the amount of pressure needed to produce a break. A nautilus shell was put between the jaws, and the type of damage inflicted on the shell was observed. For several days Tom Daniel's fourth-floor lab reverberated with cracks and snaps as shell after shell fell victim to these jaws of death. An army of mosasaurs could not have had so much fun. And in all the carnage that ensued, not once was a round circular hole approximating the size of a mosasaur's tooth ever produced.

All nautilus shells we had smashed up to this point had been empty. What might be the effect of a living body in the shell? Might a circular hole be punched in a shell filled with flesh? To test this possibility, Erica constructed nautilus "bodies" out of Jell-O and stuffed them into the shell. Once again, the shells smashed gleefully, and larger circular holes the size of the teeth never formed (Smaller holes, never circular, *did* appear rarely in the walls of the shell in regions underlain by the chambers.) A conclusion was reached: Unless ammonite shells were radically different from nautilus shells (which all investigators agree is not the case), there is no way that a mosasaur could have produced a circular tooth hole in an ammonite shell.

The mosasaur jaw produced in Tom Daniel's lab was constructed with a single intention: to test whether a circular hole could be punched in a chambered cephalopod shell—and thus to falsify (if possible) the hypothesis that mosasaur teeth left circular tooth marks in ammonite shells more than 65 million years ago. That mechanical jaw was crude to look at, but it did its job. Soon, however, a far more realistic test was made.

The Kase–Johnson team was also interested in actual tests, and they had access to far better material. The Tyrell Museum has a large number of beautifully preserved mosasaur skulls in its collections. Using molding and casting techniques, skilled technicians at the Tyrell created a hard-plastic cast of a large skull, teeth and all. Not content merely to munch nautilus *shells*, however, the masters of verisimilitude at the Tyrell did our experiments one (or several) better: They took their mosasaur head to the Philippine Islands and tried it out on living nautiluses in the sea.

I can only imagine the looks the Filipino customs agents must have given this grotesque bit of carry-on luggage. And it is a good thing that the Society for the Prevention of Cruelty to Animals didn't get wind of this little gig. The chutzpah was amazing. And who came up with the money for this? In any event, off they went, and sure enough, mosasaurs could certainly break into shells, but making little cookie cutter holes in these shells was not in their ancient repertoire.

What can we say to end this story? I have been a bit unfair to Erle. Ironically, I think Erle Kauffman was right for the wrong reasons. Did mosasaurs eat ammonites? No current evidence suggests that they did. The bite story as originally devised, and then later amended by Erle, is clearly implausible. But I happen to believe the mosasaurs *did* eat ammonites. Ammonites, after all, were among the most abundant and commonly encountered food resources in the Cretaceous oceans. It is absurd not to assume that large carnivorous vertebrates then (as now) relished cephalopod meals. The elaborately ornamented ammonite shells, especially among larger species, were probably effective designs to thwart smaller predators, such as most fish. Mosasaurs, however, could have been highly efficient ammonite eaters. The only problem—for us paleontologists—is that such attacks would *never* have left any sort of identifiable mark in the shells, other than a few cracks. Such shells, after attack, would have been no more than a few calcareous shards drifting to an uncertain fossilization at the bottom of the sea. Who would recognize such fossil fragments as the evidence of an ancient predatory attack?

Each spring, the eroding cliffs now seem to yield a new marine reptile from the black shale of Vancouver Island. Sooner or later, I am sure, one of these wondrous fossils will erode out of Sucia Island as well. It is certain that mosasaurs great in size and number once roamed the ancient seas of this region, and surely they would have been wonders to see. Now, sadly, they roam only in our imagination—and sometimes that imagination is stretched too far.

Fossils come down to us through the immensity of geologic time. They are our data. They do not change, but how we interpret them does. It is we humans who have to make the interpretations, and we are highly fallible time machines. The fossil record can be interpreted in myriad ways, and finding the truth is no simple matter. There are ammonite shells with holes in them, of that we are sure. Of the rest, we can only subject hypotheses to tests that will demonstrate the falsity of those that are disproved. But whereas hypotheses can be shown to be in error, they can never be *proved* true. That is the nature of the time machine we call the scientific method. We cannot be sure that limpets drilled the telltale holes in ammonites. But we can be confident that mosasaurs did not.

7

Virtual Ammonites

How can we reconstruct the life of an extinct animal, such as an ancient ammonite? How is it that paleontologists have revivified dinosaurs as agile and obstreperous? How can they assert that a pterodactyl could fly rather than simply glide? These and many more feats of apparent sorcery fill the scientific journals and sometimes spill over onto the wide screen and the wider popular consciousness. They derive from a type of scientific analysis called functional morphology, which is a way of studying biological structures to determine not how they looked but how they worked and what they did. This form of analysis is among the most useful of time machines.

Scientists who are interested in reconstructing the past lives of ancient organisms by studying form and function recognize three primary factors that must be considered in interpreting biological structures. First, and most obvious, are the functional considerations: Just what does this structure *do*? Most biological structures yield some obvious clue. Wings are clearly for flying, fins for swimming, legs for walking. But sometimes things get far more complicated. Some structures are *not* as we would imagine them to be to fulfill some obvious function, whereas others share a function, or perform two or more tasks, and thus become compromises in form. Take the gills of clams, for instance. Gills first evolved for respiration; they are used to extract oxygen from sea water. But in some clams, gills are also used for feeding. The same structure now has two functions. It is optimal for neither but, by doing both jobs, serves its owners well.

Other structures seem incomplete, only partially useful, or so woefully inefficient as to be laughable. Why do we have appendixes and wisdom teeth, for instance? Some structures have no obvious function at all. Interpreting task, then, is not enough to explain why some organs or biological structures have the anatomy that they show. Yet evolution, which shapes all life, rarely produces wasted organs or structures. Perhaps more than simple function is at work here.

The second factor that influences function is related to constructional aspects. Organisms have a very finite suite of material they can build with. There is no Teflon, or WD 40, or duct tape, or stainless steel, or a million other human-made building materials in the animal and plant worlds. Some biological structures have to be made of material less than optimal for their function and are "designed" (have evolved) accordingly. Our teeth, for instance. They wear out, they are made of a mineral (apatite by name) that abrades and is attacked by decay over the years. Why didn't we evolve stainless steel teeth, which would be far superior? The answer is obvious. Our bodies can secrete the mineral apatite, but they cannot produce steel.

Finally, any biological structure must be viewed from a historical perspective. Evolution is like a river, and although it can eddy, the river eventually flows in one direction. Some structures or proteins evolved by organisms early in their history stay with them throughout time and may play a large role in how they and their ancestors are eventually made. My favorite example of how historical factors affect the design of organisms comes from the study of the octopus. Octopuses are among the smartest creatures in the ocean (much more intelligent than fish, for example), and one wonders why their brains did not continue to enlarge until they, like us, became truly sentient creatures. Instead, they dwell in the nether world of semi-consciousness. In a classic account worked out by my friend Martin Wells of Cambridge University, it was shown that the octopus suffers the ill effects of a choice made half a billion years earlier by its hoary cephalopod ancestor, the first nautiloid. This creature and its ilk evolved a copper-based rather than an iron-based blood pigment to carry oxygen. Nautilus (and hence, we presume, the earliest nautiloids as well) is a quite stupid beast with a very small brain. A copper-based blood pigment carries more than enough oxygen to meet the needs of the nerve cells and brain of a nautilus. However, copper cannot carry as much oxygen in blood as iron can. With the evolution of a larger cephalopod brain, oxygen availability became an issue for the first time. Because of its large size, the poor octopus's brain is nearly always on the brink of oxygen starvation, for nerve cells above all others need a constant supply of oxygen. The octopus cannot evolve a larger brain, because its blood supply will not support more nervous tissue. Worse yet, so basic is the type of blood pigment that the cephalopods cannot redress this wrong by evolving an iron-based pigment at this time. It is too late. This wrong choice offers an example of the historical aspect of functional morphology at work.

The study of form and function is currently a growth industry in science. Although this type of study has a long history, the field has taken on a rigorous demeanor only in the last three decades. In this interval, modern engineering techniques have been employed in studying biological structures, and much of this new emphasis stems from Duke University, where (among others) Professor Steven Wainright and his colleagues and students have revolutionized the field. A considerable number of paleontologists specialize in functional morphology. Those practicing in this relatively new field often have only sketchy evidence to work with; yet they are required to understand and synthesize—literally to "flesh out the bones"—in such a way as to bring extinct creatures to life. Yet functional morphology is not restricted to those who study the fossil record. The living as well as the long dead are scrutinized, and often it is an understanding of the living relatives of extinct animals that makes possible a critical insight or breakthrough.

Perhaps the most challenging functional interpretation occurs when we address a creature that has *no* living counterparts. Many of the Burgess Shale creatures of Cambrian fame (found on a mountainside in British Columbia and vividly described by Stephen Jay Gould in *Wonderful Life*) are in this category. They often look like nothing still alive, and understanding the function of their peculiar bodies has taxed the imagination and skill of many scientists. Yet sooner or later, some insight is reached, and more often than not the critical breakthrough comes from two sources: a better understanding of similar bodies or body parts among the closest living forms (even if they are only distantly related) and analyses that borrow from tool kits employed by engineers. In fact, the discipline of functional morphology is rapidly being co-opted by engineers who have found that studying animals is much more fun than building bridges!

One of the great advantages of this type of work is that several new approaches enable us to "revive" long-extinct animals with computer technology. We can make "virtual animals" by recreating them on computer screens, using the same sorts of imaging and technology that brought the dinosaurs of Jurassic Park back to life. Dinosaurs may be long dead, but we can study their computer images. We can often achieve new insights into their biology just by seeing the reconstructed animals on the screen. And there are even more powerful applications. Frequently, parts of ancient animals are reconstructed and then subjected to various computer-simulated pressures, stresses, and strains—an imaginary animal encountering imaginary forces. Yet the results are anything but imaginary; they can lead to critical insights. Virtual modeling is a very powerful time machine.

The mystery of the ammonites

In my own field, all of these elements have recently coalesced to solve one of the longest-running mysteries in paleontology: Why did ammonites, whose closest living relative is the chambered nautilus, have such complex sutures (the intersection of their chambers and shell walls)? And why did these sutures,

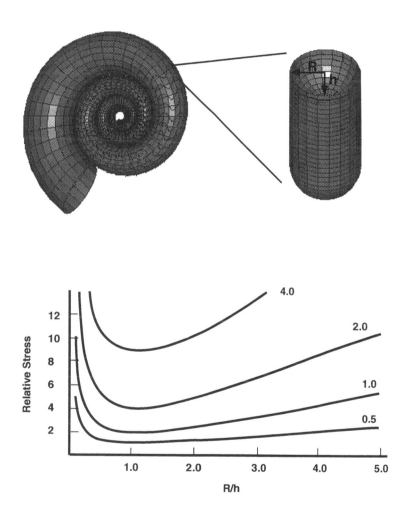

"Virtual ammonites" created by computer, with some of the data illustrating that increasingly complex sutures led to weaker shells.

which often can be used like a fingerprint to identify an individual species, get more and more complex through time? The answers to this particular problem came from an engineer.

Understanding ammonites, perhaps the most common and iconic of all Mesozoic invertebrates, really starts by studying submarines. And a horrible marine accident that happened nearly 30 years ago focused much attention on how ammonites may—or may not—have used their complex septa (the walls that divide the shell into chambers).

On April 10, 1963, the United States nuclear submarine *Thresher* exceeded a still-classified depth and imploded, killing its 146 crew members in the process. Its remains sank to the bottom of the Atlantic Ocean, at a depth of 8500 feet about 100 miles east of Cape Cod. They rest there still, along with the remains of the crew, scattered over 400 square yards of muddy bottom. The United States Navy, using remote submersibles, visits this sad grave regularly to monitor the fate of the poisonous plutonium and other radioactive waste now buried at this site. Four or five larger pieces can be observed: The sail, sonar dome, bow section, tail, and engineering sections are recognizable. The rest of the ship is represented only by scattered fragments. The pictures that emerge bespeak the violence of the implosion event itself and give mute but eloquent testimony to a law of submersibles: There exists a maximal depth below which a submarine cannot dive. If this depth is exceeded, violent, catastrophic implosion ensues.

Almost 14 years to the day after the Thresher's hideous death, I anxiously watched a winch do its slow work on the back of a Fijian fishing boat several miles off the verdant coast of Viti Levu, the largest island in the Fijian archipelago. The day before, I had caught eight healthy nautilus specimens, creatures at the time more precious (at least to a biologist) than any gold or silver. I had not counted on such a large haul and had decided to "bank" them in a closed trap. They had been caught at over 500 meters, which is about the greatest depth at which these last members of a shelled cephalopod clan can reliably be caught, and they had been returned to that depth. But this morning my depth sounder showed that the trap I had lowered to 500 meters rested now at 750 meters instead. During the night, heavy surface waves had tugged at the buoy, dragging its tethered cage below to a greater depth on the steeply tilted sea bottom.

First visible only as a faint patch of light in the deeper blue of the sea, the trap eventually metamorphosed into a large chicken wire and rebar cube. But instead of the eight living nautiluses that I had placed there for safe-keeping, I could now see only shell fragments. When we finally muscled the trap over the boat's side, the disaster was confirmed. Eight fleshy carcasses rolled lifelessly about on the bottom of the trap, already much scavenged by the countless smaller denizens of the sea. Yet it was the shells that were of most interest; it was they that told the tale. They were smashed into numerous small pieces, as though some malicious giant had whacked each with a large hammer. The pieces were priceless: They made possible a detailed analysis of how pressure caused mechanical failure in a shelled cephalopod, for the nautilus and the *Thresher* had fallen victim to the same sad fate. Both had exceeded the greatest pressure that their shell was designed to withstand.

Implosion, a terrible end for a submarine, is the handmaiden of hydrostatic pressure. This inexorable force pushes against every square inch of any submerged vessel containing gas-filled spaces, and it increases with depth. It is the force that causes ear pain in a swimming pool. It is the force that killed the *Thresher*. Such a Damoclean sword is certainly noticed by the engineers who design submarines, and it has certainly been noticed by nature's engineer, the principle of natural selection. It comes as no surprise that most interpretations of ancient submarine design—the form of the nautiluses and their close relatives, the extinct ammonites—have been viewed with this paramount fact in mind: Shell strength must be the primary concern of every one of these ancient or modern diving devices. So I was taught by the greatest student of these creatures, Professor Gerd E. G. Westermann of McMaster University in Ontario, and until recently I had no doubt about the veracity of this interpretation. Now my mind has been changed—a reversal brought about not by the study of any new fossil, living animal, or military submarine, but by experiments conducted by another investigator and his trusty Silicon Graphics workstation. The life of ancient ammonites has been resurrected by the development of a virtual ammonite.

Ammonites have interested and mystified "natural philosophers" since the Renaissance. In the 1600s, the British naturalist Robert Hooke illustrated some of the copious ammonite fossils to be found at many localities around his native Britain and made the first informed speculations about how they may have lived. He noted that these ancient shellfish had shells partitioned by many calcareous walls (the septa), which divided the shell into individual chambers. Hooke correctly deduced that the ammonites from his collections were closely related to the living chambered nautilus, known to him from shells brought from the far tropical Pacific Ocean.

Ammonites and the nautilus both live within a chambered shell, and the shell itself is cordoned off by the internal septa. It is thus reasonable to assume that in many ways they shared a similar mode of life. But Hooke recognized that these two creatures differed in one great respect: The septa of the nautilus, numbering about 30 in an adult shell, have a simple concave shape. They are like small watch glasses fitted within the shell. Ammonites, on the other hand, exhibited vastly more elaborate septa. Whereas their middles might be only slightly curved, the margins of the ammonite septa are crenulated into a complex series of curves and vaults that intersect the shell as a very complex pattern. This intersection is called the suture. In nautiluses, the suture describes a simple if slightly wavy line. In ammonites, the suture is a very complex curve—a sine wave on LSD.

Hooke was most curious about how the nautilus and ammonites might live. He thought that both animals used their shelled portions as a buoyancy device and suggested that the internal, chambered portions in both were filled with air. But air at what pressure? Hooke was a genius, a person eclipsed in fame only by his immediate scientific antecedent in Britain, Sir Isaac Newton. Hooke asked the following question and, in so doing, set the stage for one of the most common approaches in functional morphology: If this mechanism functioned as he thought, how could it best be engineered? Specifically, if some creature wished to have a gas-filled shell to use as a buoyancy device, how could it most efficiently be built and most effectively work?

The answer Hooke arrived at was simple: The ammonite would have a tiny gas gland capable of filling the chambers with air at a pressure equal to that of the surrounding water. With such a system, the ammonite would be unlike a submarine: It could descend to any depth and never worry about implosion, because its internal pressure would always match the external pressure, which increases by 14.7 pounds with every 33 feet of depth descended.

Hooke's supposition, that ammonites (and the nautilus as well) were capable of equalizing their internal gas pressure with the external pressure through the use of a gas gland, turned out to be false. Instead, these two groups of chambered cephalopods use (or used) a far less elegant solution: Their internal parts are always at *lower* pressure than ambient (the water pressure of the depth they inhabit at any given moment). In fact, studies have shown that the chamber pressure is usually *far* lower. Nautiluses (and probably ammonites) are thus susceptible to implosion when sufficient depth is reached. They therefore must depend on the strength of their shells to allow them to descend to great depths in the sea.

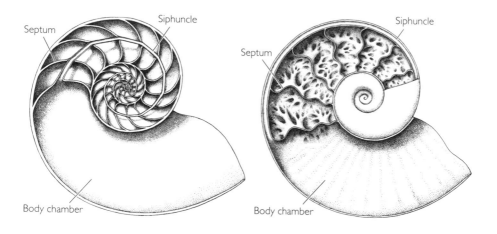

Comparison of shell interiors from nautilus (*left*) and ammonite (*right*). Note the far greater complexity of ammonite septa.

Hooke was very much a pioneer and a person of unimaginable brilliance. He was also very far ahead of his time, and he had set out a clear research program to investigate these animals further. Yet little more was written about ammonites for two centuries after Hooke's death, until these curious fossils called once more to scientists—this time as time markers, not biological curiosities.

By the mid-1800s, ammonites had taken on enormous importance to the scientific world, for, as we have seen elsewhere in this book, they were early recognized as being the most useful of all fossils for subdividing strata into time units. The pioneering biostratigraphical work of William Smith in England and D'Orbigny in France was carried out largely with collections of ammonites. These coiled fossils were regarded literally as godsends, bits of matter placed on earth by some convivial deity to help us plodding humans figure out how old rocks are. Yet gradually, keener minds began to return to the biological questions first posed by Hooke centuries before. The fossil ammonites were, after all, the remains of once-living creatures. How had these mysterious creatures lived, and above all, *why* had they produced such elaborate sutures?

For reasons quite obscure, fossil cephalopods (which include ammonites) have attracted some of the more colorful personalities ever to practice paleontology. Perhaps coincidence, perhaps more, but no other area in paleontology is peopled by a more bizarre cast of characters than the students of cephalopods, and especially those interested in ammonites. Examples are numerous and will be but briefly noted (I want to keep some friends, after all). The late Rousseau Flower comes to mind, a garrulous sort who earlier in this century collected his cephalopod fossils garbed in cowboy outfit, including two Colt pistols (he also attended a national geological society meeting wearing a gorilla suit). Then there is the late Ulrich Lehmann, a mystic and astrologer (as well as an ammonite specialist) who consulted the stars in advance of research projects. And there's even my close friend and colleague W. James Kennedy, who held a soiree attended by the president of Oxford University in honor of the birthday of one of his pet guinea pigs. But for sheer eccentricity, few ammonite workers can match the nineteenth-century British cleric and naturalist Dean William Buckland.

His grave sits in a quiet glen not far from Oxford, surrounded by the Jurassic ammonite-bearing exposures that were his life's avocation. But Buckland was no mere paleontologist. He was an ordained minister and elder of the Church of England, and indeed he was devoutly religious. He was also clearly "certifiable," as my mother would say. Even more than a century after his death, stories abound. Perhaps his most memorable act, performed while on a visit to the Natural History Museum in Paris, was reaching into a bottle that held the preserved heart of Louis, the last King of France, snatching out the dripping organ, and taking a great bite from it, shouting, "I eat the heart of the King of France!" Ah, nationalism. Lucky for Buckland the kingly heart was preserved in ethyl alcohol, rather than in methyl alcohol or formalin.

Early in life Buckland became fixated on two deities, God and ammonite fossils, and he managed to mix them together in producing a memorable nineteenth-century work known as *The Bridgewater Treatise*. Buckland thought he could prove the existence of God by showing the perfection of various aspects of nature, ammonites included. He thus set out to illustrate how ammonites may have lived.

The Bridgewater Treatise is of interest to modern ammonite workers because it presents the first detailed hypotheses on ammonite sutures since the time of Hooke (and it far surpasses Hooke's interpretations in its detail). Buckland was obsessed with ammonites during his life and saw thousands of specimens. He knew intimately the nature of the intricate suture marks lining the outer shell, and he pondered their meaning. His explanation for their presence was novel: He saw them as buttressing structures put into the shell (by God) to increase shell strength. The more complex the suture (the intersection of the chamber partition with the outer shell wall), the stronger the shell. This explanation seemed obvious to Buckland. Just as a corrugated roof is stronger than a flat sheet of metal would be, so too would corrugated ammonite septa be stronger than simpler, flatter septa, such as those possessed by the nautilus. Buckland eventually went to meet his Maker, and his vast *Bridgewater Treatise* remains a curious and somewhat derisive monument to one man's life. But among Buckland's theories, virtually all now dismissed or disproved, his comments on the function of ammonite septa in strengthening the shell stand shining apart. They were largely accepted up to the present day.

Does more complex mean stronger?

In the early 1900s new breed of paleontologist was emerging, investigators concerned not only with what fossils could say about time but with what they could reveal about themselves and the world they lived in. The analysis of function began to be taken seriously. Foremost among this new wave were Germans, most from the University town of Tubingen. Tubingen already had a long association with paleontology, having been the home of two of the greatest of all early paleontologists, Albert Quendstedt and his remarkable pupil, Albert Oppel. But these nineteenth-century paleontologists were concerned with biostratigraphy, whereas their early twentieth-century descendants thought more about mode of life, form, and function. Because they were surrounded by rocks rich in fossil ammonites, it is no surprise that many of these students began to study ammonite paleobiology.

One of the most brilliant of these students was A. Pfaff, who made the most detailed functional study of ammonites up to that time. He noted that ammonites with compressed shells had more crenulations than those with rounder shell cross sections. From this observation he deduced that the flatter shell regions were inherently weaker (because of their lack of curvature) and would therefore need more shell buttressing. Rounder shells would be supported by the curvature of the shell and would require less buttressing—and thus less complex sutures.

Pfaff and others of his time were also well aware of another curious facet of ammonite sutures: Over the long, 360-million-year history of the group, the septal sutures just seemed to keep getting more complicated. Ammonites didn't show a greater number of sutures over time—in other words, their shells did not get packed with more and more septa—but those septa that were present became more sinuous in their appearance until, by the Cretaceous Period, they made the most complicated river course seem straight by comparison. (Stephen Jay Gould has disputed this long-held view in his recent book, *Full House*. Gould quotes the work of two paleontologists who used fractals to characterize the complexity of ammonite sutures and concluded that this increasing complexity is less clear-cut than has been supposed. This may well be, for there *are* reversions toward simplicity in some lineages, most noticeably in a group of Cretaceous ammonites called the neoceratites. All in all, however, a pervasive trend toward increasing complexity clearly occurs in most evolving ammonite lineages.)

In the 1930s a new paleontologist arrived at Tubingen, and his brilliance eclipsed even that of his illustrious predecessors. Otto Schindewolf arrived at Tubingen as a young professor and proceeded to become the most influential paleontologist of his generation. Schindewolf considered many large-scale questions, such as the nature of evolution and mutation rate, the cause of mass extinctions, and, of course (like all good German paleontologists), the curious evolution and pattern of ammonite sutures. Schindewolf was so gifted that he not only attracted German students but also drew young paleontologists to his labs and lecture halls from across the seas. These disciples included young George Gaylord Simpson, an American paleontologist who was to reach the top of his profession as well.

Schindewolf kept teaching and conducting research even during World War II, and soon thereafter a succession of outstanding ammonite workers were graduated from Tubingen, including Dolph Seilacher, Jost Wiedmann, Jurgen Kullman, and the man whose name was to become synonymous with the functional morphology of ammonites and their complex suture, Gerd Westermann.

Thanks to a providential attack of jaundice, Westermann just missed being sent to the eastern front as a boy soldier. Soon after the war, he was conscripted by the victorious allies as a coal miner, but he soon escaped that fate and completed his Ph.D. under Schindewolf's supervision. With degree in hand he immigrated to Canada, where he received a university post in the early 1960s. Westermann quickly rose through the professorial ranks at his college, McMaster University in Hamilton, Ontario. Much of his early work was on the biostratigraphy of Jurassic ammonites, and like all good German paleontologists, he compiled and published massive monographic treatments on his specialty. But unlike many of his colleagues, Westermann began to wonder not only *when* his fossils lived but also *how* they lived. He began to delve deeply into the structures and possible workings of these most enigmatic of fossil invertebrates.

In 1971 Westermann published a landmark and revolutionary paper about ammonite paleobiology. By incorporating newly available information about the function of the nautilus shell based on breakthrough work conducted by Eric Denton and John Gilpin-Brown in the mid-1960s, Westermann proposed that the relative habitation depth of any ammonite could be determined by a simple measurement of two parameters found within every ammonite shell. Although the method did not reveal the actual *measured depth* at which a given species of ammonite resided, it yielded the relative depths of various ammonite taxa: that species A was capable of inhabiting deeper water than species B, and that B lived deeper than C, and so on. A corollary to this discovery was that those ammonites with the most complex sutures generally showed the deepest depth ranges. All of the paleontologists back to Buckland seemed vindicated. Ammonites with complex sutures lived deeper than those with simpler sutures.

During the 1970s Westermann kept making discoveries that painted a whole new picture of how these ancient creatures lived and "worked." He used innovations, employed a variety of engineering techniques, and in pioneering fashion even co-authored a paper with a member of the engineering faculty at his university. One of his findings was that ammonites with highly complex septa had thinner septa than those with simpler morphologies—and far thinner septa than any of the nautiloids, the supposedly most primitive forms with the simplest septa of all, watch-glass–like structures slung between the shell walls. There was only one snag in all of this: Nautilus, with the simplest of septa, managed to live in very deep water. Nevertheless, a general model seemed to be emerging. Ammonites evolved ever more complex septa that were thinner than those with less complex morphologies. Simple but strong septa would have to be very thick, and making thick septa "cost" the animals in that they had to expend more metabolic energy. An alternative was to make thin septa but septa so complex that they were stronger, than thick septa of lesser complexity.

I was privileged to be a grad student of Gerd's at the time he was making these seminal discoveries. His office was always open, and we often talked of ammonites and many other things. Why would they have produced such complex septa? Could some other function have been involved? Might the very complexity somehow have been used for a secondary function, such as some type of respiration, or muscle attachment, or even the removal of liquid from newly forming chambers? Westermann would mull these possibilities over and then reject them. Structural support seemed the only possibility.

The intricacies of ammonite septa continued to intrigue paleontologists during the 1970s and 1980s, but Gerd Westermann was king. Refinements to the model of septa as structural support were made, and Westermann, in collaboration with a brilliant post-doctoral student named Roger Hewitt, began using ever more sophisticated engineering techniques to study these forms. One generalization reigned: Ammonites with more complex septa were capable of living at greater depths in greater pressures. It was a story of imperialist evolution: ammonites invading ever deeper waters through time, colonizing regions of the sea that were previously too deep for their shells. One could envision legions of new ammonite food succumbing to the first prey—their shells, and depth, no longer protecting them from these efficient Mesozoic predators.

From Buckland to Westermann was over a century, a long time in science. Ammonites were touted as engineering marvels, miniature submarines with shells designed to withstand the ocean depths. Evolution, that supreme engineer, even kept improving the design by creating ever more complex septa. Who could disagree?

The first crack in the edifice of this wonderful paleontological story came in 1994, and it came from Bruce Saunders of Bryn Mawr College. Saunders is a specialist in Paleozoic ammonites, which lived from their inception in the Devonian Period, some 400 million years ago, until the end of the Permian Period, some 250 million years ago. The ammonites of this time possessed relatively simple septa and sutures, but they did show a progressive increase in complexity from the earliest to the latest. Saunders, who was interested in the shapes of these many early ammonites, conducted an extensive study of many taxa. He amassed a large collection and then proceeded to cut each of these specimens (some priceless, most not) in half with a diamond saw. The perfectly bisected fossils could then be easily measured for their shell shape. At the same time, Saunders decided to measure the thickness of their septa, because these were visible, and because no one had yet repeated Westermann's experiments in correlating septal thickness with septal complexity (which had been conducted only on Mesozoic ammonites). In the Mesozoic specimens he studied, Westermann had found that the ammonites with the most complex septa were also those with the thinnest. To his surprise, Saunders found no such relationship in his Paleozoic ammonoids.

It was clear to Saunders that something was amiss. Coincidentally, I was visiting Saunders in his Bryn Mawr lab soon after he made this discovery, and we talked about the implications. I agreed with him: If more elaborate septa were *not* thinner than those of lower complexity, what was the use of going to all the trouble of elaboration—unless this particular form was serving some other function. But what? Perplexed, we did what most people do when facing a problem. We sought help.

Help in this case came from Tom Daniel of the Zoology Department at the University of Washington, whom I knew slightly at the time (we met him in Chapter 6 on mosasaurs). He had come from the center of functional morphology at Duke University, where he had been a Steve Wainright student. Tom had started as an engineer and then switched to zoology. He was thus ideally trained to interpret form and function.

He worked in a cramped lab packed with wires, machines, microscopes, computers of all makes, and students. Especially students. Students of all shapes, forms, and levels. Graduate students and post-docs, undergrads and scientific visitors. Tom was not yet forty, yet he had already received the University's Outstanding Teaching Award, as well as having distinguished himself in a slew of scientific papers. He was a magnet for those who wanted to learn about how animals work. Walking into Tom's lab was like walking into a hurricane: noisy, energizing, exhilarating, chaotic.

The man himself was perched in an alcove piled with papers and electronics. Fueled by sugar, caffeine, and cigarettes, he barely touched the floor while walking. Tranquility was not part of this package. Humor was. The lab ran on laughter and respect. It emanated good feeling, and it generated good science.

"What can I do for you two boys?" Such was our introduction to Tom Daniel.

Virtual ammonites

We had brought several ammonite specimens for Tom to peruse. We explained that the complex septa had for centuries been considered structural supports but that we had our doubts. What did he think? Tom spent several minutes turning over the specimens, looking at the septa, muttering to himself, singing songs, and trying out various accents. Then he pronounced, "Well, they certainly weren't used for strengthening the shell." Thus began a 4-year collaboration that continues still.

Saunders and I convinced Tom Daniel that the "ammonite suture problem" was important. Perhaps to humor us, or perhaps because he saw the beauty and mystery in such complexity, Tom decided to take on this problem, but in an interesting fashion. He decided to test those aspects of shell strength that were related to sutural complexity not on actual specimens, and not even on models made of some bioplastic or other casting material. He decided to build what we would eventually call "virtual ammonites" on computers.

A year or so before our arrival with fossils in hand, Tom Daniel had been awarded a grant for equipping his department with Silicon Graphics workstations. The year before, these computers had gained instant notoriety when they were used by the creators of Spielberg's *Jurassic Park* to bring the film's stars—the dinosaurs—to life. Tom's acquisition of so many of these workstations had made his computing center the rival of small software corporations. Even more impressive, Tom actually knew how to use them.

But how to test the strength of an ammonite? How to test the hypothesis that increasing septal complexity made the shell stronger? One could make exact molds. But that would require many weeks of skilled artisan work, and even then it would be impossible to determine whether the artificial ammonites were responding to pressure in way faithful to the original— whether the artificial shell was actually of even strength and was failing because of material rather antique design properties. Test an actual specimen? As beautiful as many ammonite fossil are, it is doubtful that any have survived the minimum of 65 million years entombed in sediment since their last demise without undergoing mineral changes that would compromise the delicate balance of shell strength. These intractable problems had stymied all past investigators. Tom Daniel proposed taking a new tack: Build the ammonite on the computer, and then subject it to simulated pressure.

Just before our chance arrival (a visit rued by all at times over the next 4 years), Daniel had bought the site license for a new generation of engineering software program. One of the most powerful of all engineering procedures is called finite element analysis, or FEA. FEA works by subdividing complex structures into small elements and then monitoring forces that act at the junctures, or nodes, of each of these elements. It thus embodies the time-honored method of attacking a large and difficult problem one small step a time. Any structure is broken up into thousands of tiny regions, and the stress acting on each region is examined separately. A map can then be drawn to show how forces acting on the entire structure are encountered.

FEA had long been the province of those blessed with large mainframe computers. But the computer revolution of the past decade had packed computing power into very small boxes. The Silicon Graphics workstation had the processor and enough memory to construct a charging *Tyrannosaurus rex*—or to build an ammonite shell, node by node, and then subject it to pressure.

And so we began building ammonites by computer. Tom's first efforts were crude, but just as evolution starts simply and builds up, so too did the virtual ammonites become more and more elegant. And even our early efforts yielded fascinating results: As the septa in the virtual ammonites became more complex, the stresses applied to them became more catastrophic. Our first runs examined simple cup-shaped septa and then septa with folds and kinks that looked very much like the septa of early ammonites. The simple cap-shaped septa proved far more resistant to stress; this was the insight that had led Tom, on our first meeting, to suggest that ammonite septa had not evolved primarily because they afforded strength against implosion. This was clearly not the scenario that ammonite paleontologists envisioned. The evolution of more complex septa was supposed to make the shells stronger, not weaker. Perhaps it was the methodology? Were we forgetting something? We went back to the drawing board. Bruce Saunders and I were sent to our collections, where we sectioned ammonites and made exacting measurements of the thickness of wall and septa, for we needed to enter these values into the computation models if any sort of relationship with reality was to be achieved. In the process we learned much about the average thickness of real ammonites, and all this information was fed into the models.

Weeks turned to months, and months to years, as Tom and his grad student Brian Helmuth, who had joined the project, produced ever more elaborate ammonites. Computational time soared. We were now running terabits of computations, and eventually eleven Silicon Graphics machines were running simultaneously, days at a time, for each run required an hour and yielded just a single data point. Over time, the simulations were made not just on septa but on shells with septa within; the results as they flashed up on the screen were hauntingly beautiful. The images showed the distribution of stresses as colors, and the ammonites would light up as spectacles of sparkling reds and yellows, blues and greens dancing across septal junctions and shell wall. By late 1996 an entire school of ammonites lived in the basement of Kincaid Hall, populating the bowels of the building rather than the ocean deeps. The machines now ran simultaneously, and Brian Helmuth ran between them for all the hours of many days, plucking a new number from each laborious, hour-long computation. As *T. rexes* lived and died in the California studios making *Jurassic Park*, the brethren of these powerful machines raised ammonites from another grave in Seattle.

All this effort required money. Grad students don't come free, and computer time costs. Early in our efforts, we sent a proposal to the National Science Foundation soliciting a modest grant to fund our study. Such grants are peer-reviewed; about a dozen colleagues read each proposal (a document about fifteen pages long) and rate it anywhere from Excellent to Poor on the basis of the merit of the proposal, its significance, its practicality, and the qualifications of the proposers. These grants are a bear to get—only about 15% are funded in any given year—so the competition is fierce. We submitted out proposal with high hopes, because we had demonstrated (we thought) the feasibility of our research program. Out went the grant, and six months later it came back, rejected. The reviews were split. About half of our judges were absolutely ecstatic at this new approach to an old problem. The other half were equally certain that we were completely off base, that it could never work, and that everybody knew what ammonite septa were for anyway. The low ratings doomed the proposal.

Rejection in hand, we rewrote the proposal. Another six months went by, and meanwhile the computer churned out numbers. Again a phone call, again a rejection. But the section head at the agency within the National Science Foundation to which we were submitting, a paleontologist named Chris Maples, saw merit in our approach, and threw in a small, one-year allowance that kept our machines rolling. He told us, however, that the ammonite workers of the world seemed unanimous in their condemnation. Leading this charge was none other than my research supervisor from Mc-Master University, Gerd Westermann. In several apoplectic phone calls and letters, he let me know that I must have gone mad or suddenly stupid.

As the model matured, we were able to ask new things. We built in septa of many types and designs, and we varied shell wall thickness and septal thickness as well as septal complication and design. We varied the spacing of the septa. Eventually we were able to subject the computer models to attack by virtual predators and to vary the simulated attacks. More time crept by; Tom's two infant daughters became little girls; Tom quit smoking. And then in the spring of 1996, the most wonderful thing happened: Tom received a phone call announcing that he had been chosen to receive a

McArthur Award, one of the so-called genius grants. The ammonite work gained new credibility.

By early 1997 we had enough numbers to produce a solid paper stating what we now know: Ammonite septa did get more complicated through time, but that complication had nothing to do with habitation in ever greater depth. Ironically, the earliest of all septa designs, that used in the Paleozoic Era of 500 million years ago by the earliest nautiloid cephalopods (and a plan close to the design used today by the still-living nautilus), is by far the best design for shell strength. But ammonites, throughout their history, had freely abandoned that design. Why? After all of this time and effort we knew what ammonite septa did *not* do, but we still didn't know what they *did* accomplish with their amazing complexity.

Our best guess about the real function of ammonite septa comes from studying the still-living nautilus. The nautilus achieves nearly prefect neutral buoyancy in the ocean with its chambered shell. As the animal grows, it secretes new chambers, and in these lie pools of liquid that must be removed. I have long thought that the rate at which this liquid is removed is the ultimate rate-limiting step in the growth of a nautilus, as well as in that of an ammonite. But a new possibility has been entertained in the 1990s: Perhaps it is not the rate at which a chamber is *emptied* but rather the rate at which it can be *refilled*.

The buoyancy system in these creatures—the extant as well as the extinct—is a two-way street. Not only can a chambered cephalopod get lighter in the sea (and thus reduce its buoyancy), but it can also get heavier, by readmitting liquid into previously emptied chambers. Why do this? That answer is simple. The Achilles heel of the shelled cephalopod design is the fact that *any* removal of shell at its aperture—through shell breakage by accidental contact or, more frequently, by predatory attack—makes the cephalopod suddenly more buoyant. This is probably the most dangerous thing that can happen to one of these creatures. When suddenly made buoyant through shell loss, a chambered cephalopod loses its freedom of movement. Sudden buoyancy causes it to float to the surface of the sea, where it is the helpless victim not only of creatures *in* the sea but of creatures in the air as well, the ubiquitous birds watchfully waiting for any type of food to emerge from the depths.

Sudden loss of buoyancy is certainly fatal to a nautilus. In most places where it lives, the surface of the sea is simply too warm for a nautilus to live. Surely, there must be numerous adaptations—such as rapid chamber refilling—to compensate for this hazard. But there are not! The rate of refilling in a nautilus is very slow. And that characteristic, which I discovered in the mid 1980s, may be a clue to why ammonites evolved more complicated sutures.

Tom Daniel has looked at this problem and has concluded that the complications of septa within the chambered cephalopod shell would favor the *readmittance* of water to an already emptied chamber. Because of surface tension properties, such complications actually make it easier for an ammonite to refill chambers. Because any bite by a predator would cause a sudden *increase* in buoyancy, this type of adaptation would be selected for. It is a simple and elegant solution. Ammonites with more complicated sutures would be better able to refill chambers rapidly—and thus become less buoyant—following a predatory attack. As ever more predators evolved in the Mesozoic world, including a whole army of new types of shell-breaking predators, ammonites "retaliated" by evolving ever more elegant and sophisticated ways of compensating for sudden buoyancy. If our view is correct, that meant more complicated sutures.

Microcomputers have thus transported us back into Mesozoic oceans. They have served as powerful time machines.

Where are they now, these elegant ammonites? Could not a few, at least, still grace our planet? Sadly, I know the answer to that question. They are gone, long gone, never to be resurrected, driven into extinction by the collision of a giant comet with the earth 65 million years ago. But some vestige of them persists in the fertile imaginations of a few paleontologists, for surely I am not alone in contemplating deep, calm bottoms where these ancient creatures evolved structures whose startling regularity and grace so long haunted my dreams.

8

The Ancestry
of the Nautilus

Long ago, when animals had but a single common name, natural philosophers speculated about the separateness of animals and plants. Surely the in-
dividuality of what we now recognize as species (as well as the richness and
diversity of these entities) was there for any to see. But just as surely there
were *connections* between certain organisms. Dogs are not wolves, and lions
are not lynx, yet who would deny that some connection exists? But how
much, and why? If life is a great tree, how are its branches differentiated, and
how can those pruned by extinction be detected?

The classification of life

We know so much more now. We know about evolution, and genetics, and the double helix of life uniting all living organisms on this planet. Yet there is speculation and mystery still about the categories into which we classify life (the province the field of study known as taxonomy), their validity, and the relationships among them (the province of systematics). We know that life is organized into basic units, the species, and that species somehow come into being, exist for a period of time, and then disappear through extinction. Species were once recognizable only for their morphology and then, as life scientists peered closer, through the recognition that species are entities that can successfully interbreed. Finally, as technology advanced much further, scientists delved into the very DNA of a species, to discover the nature of the unique aspects of the genome that dictate the nature of all other characteristics, be they physical or behavioral.

Defining species now dead is much harder. There is no DNA to play with, no opportunity to observe breeding habits. Assigning *fossils* to species groups is thus much more difficult than categorizing living creatures. Yet is it done—with mistakes to be sure—but done nevertheless, probably with good success. It is in the nature of *our* species to categorize things, to classify; and fossils are no exception.

Many species still living have left no fossils; usually these are the creatures with no skeletal or other hard parts that might readily fossilize. Conversely, many fossil species have no living relatives; such victims of extinction have left no kin, and these poor orphans are often left dangling in our classification schemes. But a good many animals and plants on earth today *do* have a fossil record, and it is from these organisms that we can learn the most about evolution.

No paleontologist is immune to the call of evolutionary study. Even we stratigraphic paleontologists, who comb the rock record for adequate time markers, quickly hear the siren song of evolutionary change as we dig fossils from rocks. And the next natural step for any scientist is to write about it, the evidence of evolution—to describe, to study it. That is what we do. It is certainly no burden, and is usually a pleasure, to make such studies. The fossils of Sucia are an example.

Ammonites are the greatest fossil treasures to come from Sucia. Yet here and there one can discover fossils easily mistaken for ammonites, until one looks a bit closer and sees that these particular shells are simpler, less elegant. Their chamber edges form simple wavy lines, rather than the flowery contacts so characteristic of the ammonite design. They are fossils of nautiloid cephalopods—a group represented today by the chambered nautilus—cousins of the ammonites.

Although the nautilus has been studied for centuries, many questions about it remain unanswered. Is the nautilus a very ancient creature or a very new one? Is it a living fossil (a designation first proposed by Charles Darwin for organisms with long fossil records bereft of evolutionary change)? Is it the last of its kind or one of the first? Does it have many relatives or very few? And perhaps most important, why should anyone *care* about these odd animals? For me, at least, the hook is that this endangered body plan, so rare among swimming animals in today's seas, was once clearly the rule, not the exception.

The nautilus has two great claims to fame. First, because its shell so closely resembles that of the ammonites, it can be used to understand better the complex workings of the intricate ammonite shell, which must have evolved as a compromise between protection and a form of buoyancy control. Studying still-living creatures as a means of understanding the past is one of the most powerful of all time machines, and in this case, the nautilus seems to give us an accurate glimpse far back into time.

Second, as I hope to show in this chapter, the nautilus does appear to be a living fossil, a form that has existed for millions or even tens of millions of years without change. We have discovered that some fossil nautiloids closely resemble the living species. This realization has come about only because detailed studies of the nautiloids show the evolutionary relationships of the distinct species. The classification and systematic zoology of the nautiloids have enabled us to identify them as living fossils. The way in which we study the evolution of animals and plants—including the nautiloid cephalopods—is through the disciplines of systematics and taxonomy. Combined, they are another powerful type of time machine.

The story of how the nautilus has been classified is an example of how these particular time machines operate. It is detective work, and as the work progresses and the family tree is discerned, one feels great satisfaction, for any view of a family tree always brings new knowledge and a new perspective on the past.

Imagine that you *did* have access to a time machine capable of whisking you (along with a good set of scuba gear) back to the last days of the Mesozoic Era—to 65 million years ago, just before the asteroid strike that was to end the Age of Dinosaurs. What would you see? On land, of course, you would sooner or later encounter dinosaurs in great variety and profusion. But in the sea, would it be as immediately apparent that you were *not* in our time, that you were indeed back in the Mesozoic Era? Actually, as you settled onto a Cretaceous sea bottom 50 feet deep, it might be even *more* apparent that you had arrived in a foreign, long-ago place, for in any Cretaceous ocean you would quickly encounter a fantastic assemblage of large, alien-looking shelled cephalopod mollusks acting and swimming like fish. By the Mesozoic Era, the chambered cephalopods were already ancient. As the first large, swimming carnivores in the world's oceans, they have been among the dominant marine predators for much of the last 500 million years; there are over 10,000 known fossil species. Ranging in shell size from less than an inch to more than 12 feet across, and in shape from coiled to straight to elaborate, candy-cane contraptions, the great armored dreadnoughts of the Mesozoic world that you would see would be of two kinds: ammonites and nautiloids. The former, by far the more abundant and diverse of the two groups, would be easily distinguishable by their more ornate shells and peculiar shapes. But the latter, more conservative group, bereft of the rococo ribs and spines, keels, and knobs that characterize the ammonite lineage, would look quite familiar to most of us. Who has not seen the shell of the nautilus?

Following the 65-million-year-old Chicxulub comet impact (the "K/T" extinction), all of the ammonites (like the dinosaurs on land) were quickly exterminated. Yet there were nautiloids that survived this great

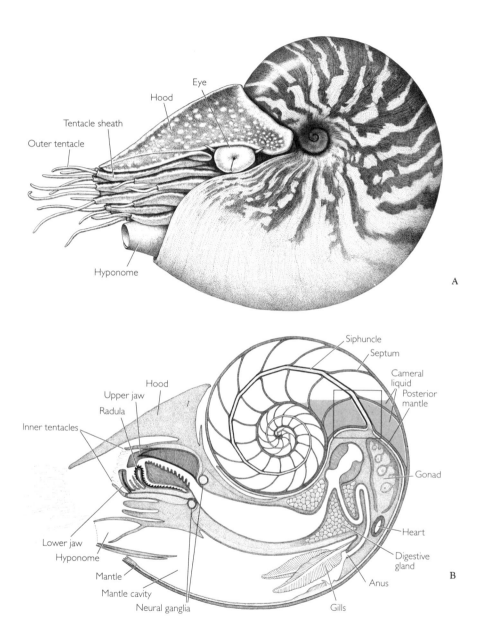

A) Exterior and B) interior views of the Chambered Nautilus. (With permission of Scientific American.)

global catastrophe. But which ones, and how many? Was it some distant *ancestor* of our familiar chambered nautilus (technically *Nautilus pompilius*)? Or was it the very same species that is living today, whose shells can be purchased in any curio or shell shop? And did only a single nautiloid survive, or might several types (either several species or higher taxonomic groupings, such as several genera or even several families) have dodged the bullet of this mass extinction? Even several years ago, paleontologists who study the fossil record of shelled cephalopods (which include the ammonites and nautiloids) would have told you that the nautiloid cephalopods living in the Cretaceous were only distant ancestors of the *Nautilus* of our world, in the same way that early apes are related to humans.

According to this evolutionary hypothesis, the single genus *Nautilus*, with its four or five living species, is the sole surviving remnant of this ancient group. Increasingly, however, this time-honored view of cephalopod evolution appears to be a hypothesis that must be rejected. New methodologies (such as advances in DNA sequencing techniques and a new way of tracking the course of evolution called *phylogenetic systematics* (also known as cladism, it is described in more detail below) and recent discoveries of previously unknown living and fossil nautiloid specimens have drastically changed our understanding of these ancient animals. Our current knowledge would have been paleontological heresy even a decade ago: The nautilus not only dates back to (and lived through) the great K/T calamity but also became the rootstock for an evolutionary radiation of new nautiloid genera in the ensuing Cenozoic Era. And (heresy of heresies!) our world is graced with not one but *two* genera of living nautiloid cephalopods. How we arrived at these new understandings is perhaps as fascinating as the animal itself, for the recent elucidation of the nautilus's great feat of survival—it has lasted at least 100, and perhaps 200, *million* years—is a saga of how science often works. It frequently progresses in fits and starts, is shaped the influence of forceful personalities and chance discoveries, and is hampered by many decades of mistaken identity.

The pearly nautilus, with its spiraled, chambered shell, is often grouped with the coelacanth, the opossum, and the dragonfly as living fossils—archaic creatures that have somehow survived to the present. In Charles

Darwin's time, the nautilus was certainly accepted as a living fossil. But during the last century, our definition of what is (and what isn't) a living fossil has changed: We now regard them not only as organisms of great antiquity but also as belonging to evolutionary groups that, through time, have produced very few species. This is a subtle but important difference from the earlier definition. Some facets of a living fossil's habitat, or habits, not only preserve it from extinction but also somehow inhibit the process that leads to new species formation. Nautilus was struck from the rolls of bona fide living fossils when paleontologists realized that in both of these attributes it differed very much from the other recognized living fossils. Genera of nautiloid cephalopods (the group to which nautilus belongs) typically contain numerous species, not few; and most troubling of all, the genus *Nautilus* seemed not to be of great antiquity at all, but rather a recently evolved taxon.

A living fossil?

My scientific studies of the nautilus, whose earliest relatives are found in Cambrian-aged rocks only slightly younger than those in the storied Burgess Shale of Canada, dated back to 1975, when I took time out from my work on Vancouver Island to spend three months living on the tropical Western Pacific islands of New Caledonia and Fiji studying these odd creatures. My motive at that time was simple: The life of the living nautilus provided the best time machine with which to extract ammonites from their ancient past. I wanted to bring the Vancouver Island Cretaceous back to life. And to do that, I needed to study the living as well as the dead.

I had collected many nautilus fossils from Sucia Island and elsewhere. I was familiar with the living nautilus's shell, of course, but like most people had never seen the animal that lives inside, because nautiluses occupy deep-water habitats of the western Pacific Ocean. I had supposed that the living animals would be more snail-like than anything else; their shells resemble nothing so much as large snails of some peculiar ancestry. But during those three wonderful months so long ago, I learned my mistake many times over as I observed the habits of living nautiluses in their natural habitat on the seaward edges of New Caledonia. With members of a local French oceano-graphic institution, ORSTOM, I would dive the magnificent outer barrier reefs of this island after sundown, and I was routinely rewarded with sight-ings of these beautiful animals as they concluded their long nocturnal as-cents from their deep daytime habitats.

The New Caledonian barrier reefs parallel the Great Barrier Reef of Aus-tralia and, like the latter, form great vertical walls extending from the warm sur-face to cold, thousand-foot muddy bottoms. During the day, the nautiluses lurk on the deep bottoms, out of sight and danger from the efficient shell-breaking predators of the reefs' shallower regions. But at night, emboldened, they swim up the reef walls to the shallows, where they feed under cover of darkness. Like hot air balloons, they would rise upward along the reef walls, and night after night I would wait for them, hanging weightless in hundred-foot depths, sweep-ing the black water with my dive light to catch a glimpse back into time. Sooner or later an isolated specimen would swim into view—they always travel alone—and I could follow it on moonlit nights into shallower water without lights. Nautiluses swim vigorously; they are nobody's placid snail.

My initial work was entirely concerned with how a nautilus uses its chambered shell to achieve and then maintain neutral buoyancy in the sea. At the time I had no research interest in the evolutionary history of this an-imal, and for good reason: *That* story had long been worked out to virtually every biologist's satisfaction. Just as no physicist would bother recalculating the speed of light, nor any biologist again attempt to show that DNA is the molecular basis of heredity, so too was working out the evolutionary history of the extinct and extant nautiloid cephalopods, at least as understood in the 1970s, viewed as a job already done. Two giants of American paleontology had said all there was to say on the phylogeny of the nautiloids—and of *Nau-tilus*—in the 1950s. Or so we all thought.

Although the name *Nautilus* was applied to both fossil and living taxa during the nineteenth and early twentieth centuries, by the 1950s the name was restricted to the living species, even though it was recognized that many fossil forms were morphologically similar to the living species. This change in concept was brought about largely through the landmark work of A. K.

Miller of the University of Iowa, then the reigning expert on nautiloid cephalopods. Miller, a workaholic of amazing endurance and longevity, had the interesting habit of airbrushing the photos of his nautiloid specimens to make them look better in his published monographs. In spite of such idio-syncrasies (today considered anathema in all scholarly publications), he was unchallenged in his specialty. He was a champion of distinguishing taxa on the basis of only one or two morphological characters. For instance, in his most important work, published in 1947, he wrote,

> In general, differences in suture pattern [a structure formed by the intersection of the shell wall and the septum that is visible only when the shell wall is removed] and ornamentation must be given far more weight than shape of the conch in the determination of the taxonomic position of a genus or species of nautiloid cephalopods. . . .

Miller's subsequent description of the genus *Nautilus* was indeed based largely on the morphology of the suture pattern. He also recognized *only* living species as members of genus *Nautilus*. These two concepts—that nautiloid genera (including *Nautilus*) were best differentiated by sutures and ornament, and that the genus *Nautilus* had no fossil record—were subsequently adopted by Miller's heir to the title of world authority on the nautiloids, Bernard Kummel of Harvard University, who, like Miller, believed that *Nautilus* first evolved in the Late Tertiary (perhaps 5 million years ago) and was without any fossil record. In 1956 in a large monograph dealing with every known nautiloid species from the Jurassic Period to the modern day, Kummel noted, "No fossil species are assigned to the genus *Nautilus*. No Pliocene or Pleistocene nautiloids are known." This view made *Nautilus* anything but a living fossil.

Miller and Kummel also framed a standard species concept for fossil nautiloids. They believed that nautiloid genera were highly prone to producing new species. By Miller's time, over a thousand post-Triassic-aged nautiloid species had been recognized by paleontologists. Either there were an extraordinary number of these creatures swimming the world's oceans through time, or the species concept for nautiloids was exceedingly narrow.

Until recently, the only assertion that *Nautilus* might be more ancient than Miller and Kummel believed came from a Russian named V. Shimansky, who in 1957 published the description of a nautiloid fossil from rocks 30 to 40 million years old in Kazakhstan. He named this specimen *Nautilus praepompilius* n.sp. Because it was remarkably similar in shell morphology to *Nautilus pompilius*, Shimansky suggested that the genus *Nautilus* might be traced back to the mid-Tertiary, rather than to the later Tertiary or Pleistocene. But Shimansky's new species was known only from a single specimen, so it was largely ignored by subsequent workers.

Miller and Kummel's claim that *Nautilus* was very recently evolved, and therefore had no fossil record, was endorsed by each new generation of paleontologists interested in cephalopods, including J. Wiedmann in the 1960s, J. Dzik in the 1970s, and C. Teichert and T. Matsumoto in the 1980s. The latter two had a *century* of combined experience studying nautiloids by the time, in 1987, when they wrote an article on nautiloid systematics and evolution reaffirming that *Nautilus*, though perhaps somewhat older than Miller and Kummel imagined, was the last-evolved externally shelled cephalopod, was recently evolved, and was the only such creature left in the world today. Who could argue with this *Who's Who* among the greatest paleontologists of the twentieth century?

A new scheme of classification

All taxonomists of this period (including, of course, those working on cephalopods) used a similar scheme of classification: Organisms were grouped together into *taxa* because they shared specific morphologies. Opinions on *which* characters were most important in this endeavor, however, were based on each worker's experience and observations, so they differed greatly among scientists. "Good" systematic biologists were those who had a "feel" for defining important characters. This method, called *phenetics*, was originated and used by such pioneering eighteenth- and nineteenth-century taxonomists as Linnaeus, Cuvier, and Owen, and it persisted through the first half of the twentieth century. By the 1970s, however, a contrasting philosophy and methodology for discerning evolutionary lineages began to compete with phenetics. This new technique was known as *phylogenetic systematics*, or, as it came to be called, cladism.

The beauty of cladistics is that, when correctly applied, it is far less subjective than phenetic methods. Instead of focusing on one important character, or at best on a few (what is "important" being determined by the practitioner), a cladist tries to use multiple characters, thus eliminating the need for deciding subjectively which characters, or traits, should be chosen. Furthermore, it is not the shared presence of specific characters that is used to define taxa but rather the number of shared "derived" characters—that is, characters that have undergone evolutionary transformation from more primitive states. For example, all vertebrate animals (such as humans) have forelimbs of some sort. This character has been "derived" into wings in birds. Because all birds share this derived character (as well as sharing a greater number of other derived characters than they have in common with any other group of animals), they are grouped together into what we call a clade. Thus, whereas both pheneticists and cladists search for and "use" characters, a cladist gives each character at least two states: primitive and derived. At last, a new and objective philosophy of classification could be employed. Its availability demanded that every group, every past classification, be re-examined. Nautiloid cephalopods, past and present, were no exception.

As the cladistic revolution was sweeping through paleontology and systematic biology in the early 1980s, I was moving westward across the Pacific Ocean, island group by island group, still primarily studying buoyancy in the *Nautilus*. But the great change in how we classified organisms brought about a shift in my interests from functional morphology to evolution, and as I saw more and more nautilus specimens, and thus was able to observe variability both within and between isolated nautilus populations, I began to doubt the then-accepted phenetics-based classifications of the living and extinct nautiloids. Most extinct nautiloid species were known from only a handful of representatives—or even from a single specimen. Character variability within a given species was thought to be very low, but because so few specimens were typically known, this interpretation was still unsubstantiated.

Nevertheless, cephalopod paleontologists such as Miller and Kummel had used even the slightest morphological difference in suture, shell shape, or ornament as an excuse to declare new species. Adding to this was the fact that status among paleontologists was often accorded to those who "discovered" the most new species during a career. There was thus incentive (promotion, pay, prestige) to announce the discovery of as many new species as possible. Not surprisingly, "new" species proliferated in the literature.

By the early 1980s I was not alone in harboring a deep suspicion of currently accepted nautiloid taxonomy, for by this time another paleontologist interested in cephalopods, Dr. Bruce Saunders of Bryn Mawr College, had also spent several years studying living nautilus populations. He too had doubts about the classification of *Nautilus*. In 1983 we joined forces to try to understand these creatures' evolutionary history. Our first task was to discover exactly how many species of the genus *Nautilus* exist today.

Classifying the nautilus

The number of living *Nautilus* species has long been in dispute. By the mid-1800s, four species were commonly recognized: the widespread *Nautilus pompilius*, *N. macromphalus* from New Caledonia, *N. stenomphalus* from the Great Barrier Reef of Australia, and New Guinea's *N. scrobiculatus*, also known as the King Nautilus. Remarkably little was known about these four species, because only two (*N. pompilius* and *N. macromphalus*) had ever been seen alive. The other two accepted species, *N. stenomphalus* and *N. scrobiculatus*, had (like all fossil nautiloids) been defined on shell characters alone; they were known only from drift shells without soft parts. By this century, many more nautilus shells had been collected from isolated island groups across the Pacific Ocean, and, given the philosophy of classification then dominant, it is not surprising that many more new species were defined, including *N. repertus*, *N. alumnus*, *N. perforatus*, *N. ambiguous*, and, most recently, *N. belauensis*. Yet all save the last were known only from shells, and their supposed differentiation as distinct species was based only on perceived differences in shell shape or size. It seemed to many cephalopod specialists that a tremendous diversity of living *Nautilus* species existed. But one nagging problem remained: In the three species from which soft parts as well as hard parts were known, *the soft parts were absolutely identical in anatomy.* And on the basis of the great variability in mature shell size, shell ornament, and shell geometry among the various "species" observed by Saunders and myself, it seemed doubtful that all of these defined nautilus species could be valid.

Clearly, we needed more information not just from the hard parts, but from all anatomical characters, of the living *Nautilus* species. It was thus imperative to capture living specimens—and especially those forms from which no soft-part anatomy was yet known. Of all of the "species" defined only by shells, we were most interested in observing a living "King Nautilus," *N. scrobiculatus.* Unlike the other *Nautilus* "species," which differ only slightly in shell shape, this latter has a shell radically different from any other nautilus. For instance, its coiling configuration is such that it has a very large, central depression in the shell (the umbilicus). It also exhibits a square (rather than rounded) cross section and a peculiar, cross-hatch shell sculpture unique among living nautiloids. Yet in spite of hundreds of its shells having been found, no living specimen—and thus no example of its soft parts—was known. Only a single tantalizing clue existed: At the turn of the last century, an English zoologist named Arthur Willey had found a rotting carcass within the shell of a beach-stranded *N. scrobiculatus* on a remote New Guinea beach. The soft parts appeared quite distinct from those typical of any of the other living *Nautilus* "species." But the advanced state of decomposition precluded obtaining any definitive answer. It would be more than 80 years (and several generations of cephalopod specialists later) before this most enigmatic of living nautiloids was finally seen alive.

The finding of the coelacanth, thought to have been extinct for over 100 million years at the time of its capture, surely ranks as the foremost discovery of a living fossil in this century, and the recovery soon thereafter of a small mollusk named *Neopilina* (considered to have been extinct since the Paleozoic Era) was exciting as well. But a third discovery, the first capture of a living *Nautilus scrobiculatus* by Bruce Saunders, must be reckoned with these others for many reasons. The other two discoveries occurred by sheer chance; Saunders's discovery was a triumph of ingenuity and intellectual sleuthing.

Saunders at least had the benefit of knowing that his animal existed. But his rapid discovery first of its habitat site and then of the living animal itself, was extraordinary. Arthur Willey searched for this animal for two years and never found it. Within a week of arriving in New Guinea in 1983, Saunders had captured this prized creature. And what a creature it was.

The shelled cephalopod that Saunders captured in the deep blue sea off Manus Island, had a completely unexpected appearance, for it looked almost nothing like the nautiluses we had grown so familiar with. And it was not alone, for Saunders began catching both *Nautilus pompilius* and the "King Nautilus" at the same locality. Saunders discovered not only that two completely different types of nautiloids lived on earth but also that they lived side by side in the same habitat.

I was able to see these extraordinary creatures in June of 1984 when I traveled to New Guinea with Bruce Saunders. Were able to catch more of these odd nautiloids. Their shells were quadrate, with weak yellow stripes, and the soft parts looked quite unlike those of a nautilus, for the upper body was covered with thick, fleshy tubercles that are lacking in all other *Nautilus* species. But the most striking feature was the shell's cover. A thick, shaggy, orange fur covered the shell, almost completely obscuring the shell ornament and making the creature look like it wore a fur coat. Thus was my first view of *Nautilus scrobiculatus*, also known as the King Nautilus.

Over time we trapped more of them and then released them back into the sea to swim with them. They looked and *behaved* unlike any of the other *Nautilus* species we were familiar with. Their shells also bore an uncanny resemblance to one of the most ancient (and the most common) of Mesozoic nautiloids, a genus named *Cenoceras*. A heretical thought crept in to our discussions: Could this living animal, for almost two centuries placed in the genus *Nautilus*, actually belong to a new genus? Or (an even more exciting prospect), could it be an undiscovered living fossil that had existed on earth, without leaving any fossil trace, since the Jurassic Period? This question could be answered only by looking in more detail both at the fossils and at modern nautiloids.

In late 1994, Saunders and I finally got around to looking at the soft-part anatomy of *Nautilus scrobiculatus*. We performed a series of parallel dissections on the soft parts of the King Nautilus and *Nautilus pompilius*.

The time-honored method of comparative anatomy is one of the most powerful methods applied in elucidating evolutionary relationships and thus is a "time machine" in its own right. Our work showed that *Nautilus scrobiculatus*, the so-called King Nautilus, was distinct in many anatomical characters, such as gill and reproductive system morphology, shell ultrastructure, and morphology of the hood. Thus in soft-part as well as hard-part anatomical features, it was clearly distinct. But was it ancestral, or was it a descendant of some *Nautilus* species, or was it completely unrelated? To address such evolutionary issues, we needed to perform a cladistic analysis, and to do *that*, we needed to know which characters were primitive and which derived. Unfortunately, we still did not have enough information to make such judgments. Hence we turned to a new source of evolutionary evidence: the genes of the nautiloids themselves.

By the 1980s, extremely powerful techniques based on DNA sequencing afforded investigators new ways of testing phylogenetic hypotheses such as those we had framed for the living nautiloids. The power of the DNA methodology is that, unlike the traditional nautiloid taxonomy pioneered by Miller and Kummel, it looks at many individual characters (the genetic arrangements) rather than only a few (the shell and suture characters) and thus allows the use of the new cladistic approach. But for these tests, tissues from live animals had to be obtained, frozen immediately at very low temperatures, and kept frozen until the laboratory analyses were obtained. This posed huge problems for those of us collecting samples from living nautiloids. These animals lived in remote, tropical localities. In many such places there was no possibility of obtaining dry ice, the substance used to freeze and then maintain the tissues. The logistics of freezing the samples and then getting them back to North America from distant tropical regions was daunting. It took us 5 years.

Tissue samples were ultimately obtained from 13 geographically distinct nautiloid populations among the "species" *Nautilus pompilius, N. macromphalus, N. stenomphalus, N. belauensis,* and *N. scrobiculatus* from numerous localities in Fiji, Samoa, Australia, New Guinea, the Philippines, Palau, and New Caledonia. These tissue samples were analyzed first in the lab of David Woodruff of the University of California at San Diego, using a technique called gel electrophoretics, and later, using more powerful DNA sequencing techniques, by geneticist Charles Wray at the American Museum of Natural History. Both sets of analyses yielded the same surprising result: Only two distinct groups emerged. One group was composed of *Nautilus scrobiculatus*, which (according to the molecular "clock" approach of analyzing genetic distance, which compares differences in DNA results to an estimated rate of change in all DNA molecules) appears to have descended from an unknown *Nautilus* species sometime in the Miocene Epoch, or about 15 million years ago. The other group was composed of all other *Nautilus* "species." "*Nautilus*" *scrobiculatus* is so different in its gene sequences from *Nautilus* that David Woodruff suggested to me that it might not even belong in the same *family* as *Nautilus*, let alone the same genus.

The results suggested that not only did the many isolated populations of *N. pompilius* all fall into the same species (which was at least comforting) but so too did *N. belauensis, N. stenomphalus,* and (to a slightly lesser extent) *N. macromphalus*. The genetic results suggested that separating out these latter as distinct species of *Nautilus* made little biological sense, and indeed Saunders and I provided independent confirmation of this with our discovery that *Nautilus stenomphalus*, which we captured for the first time alive in 1986, was apparently breeding with *N. pompilius* on the Great Barrier Reef of Australia.

If the genetic evidence was to be believed, then the long-accepted classification of the living species assigned to *Nautilus* had crumbled. Not only did *N. scrobiculatus* appear to represent a different genus, but the slight differences in shell morphology of the other species classically placed in *Nautilus* appeared to be of little or no taxonomic significance. We had gone from five species belonging to one genus, to two genera, each with one species. Instead of being composed of many species, *Nautilus* was beginning to look much more like a typical, low-diversity living fossil. In 1995 we named the other living nautiloid genus *Allonautilus* (*allo* means "other").

That was all very well, but now we were in a quandary. It appeared that shell characters and genetic characters were giving two very different classifications for the *Nautilus* species. And because the shell characters used originally to define the living *Nautilus* species by the classical taxonomists Linnaeus, Sowerby, and Lightfoot in the eighteenth and nineteenth centuries were the same ones used by Miller and Kummel in the twentieth century for classifying fossil nautiloid species and genera, it appeared that if the modern-day species were oversplit (too narrowly defined), so too might be the thousands of fossil nautiloid species. But could the genetic results be trusted? Didn't the combined paleontological wisdom of so many past students of the fossil cephalopods mean more than a few genes? To answer these questions, we had to examine the fossil record, as well as genes and living animals, using cladistics.

Cladistics to the rescue

Bruce Saunders and I thus embarked on a new attempt to reconstruct the phylogeny of all nautiloid genera since the Jurassic Period. Our interest was stimulated equally by our anatomical research on *Nautilus scrobiculatus* and by the newly emerging DNA work conducted on tissues that the two of us had obtained in the 1980s. But a third source of evidence also was emerging, for many new discoveries of fossil nautiloids became available in the early 1990s as well.

We intended to use cladistics to analyze the various genera and species of nautiluses both living and dead. Our nearly two decades of studying *Nautilus* species had given us new insight into just what a fossil "species" ought to encompass in terms of variability; we thus used the modern animals as a guide to interpreting fossil taxa. But we needed more than simply an understanding of variability to conduct such a study: We needed many additional hard-part characters that could be recognized in fossils. The traditional shell characters, such as shape, ornament, and suture type, were useful but too few in number to allow meaningful cladistic analysis. Luckily, just at this time a wealth of morphological characters, buried deep in the middle of every nautiloid, was discovered through the detailed work of Neil Landman of the American Museum of Natural History.

The nautilus has long been known to hatch at a very large size: It emerges from its egg with seven fully formed chambers and a shell diameter of over an inch, which makes it the largest invertebrate, at hatching, in the world. Indeed, it may have been this trait that enabled it to survive the great Cretaceous mass extinction, for the nautilus appears to lay these eggs in very deep water where they take a year to hatch, and it may have been the deep-living juveniles, or even the unhatched eggs, that survived in this refuge. When a young nautilus emerges from its egg, a distinct mark is left on its shell. Landman began dissecting many types of fossil nautiloids to see whether similar marks were found in ancient species as well. He discovered not only that these marks did occur but also that many *other* types of characters could be found in the hatching stages of all nautiloid fossils.

By combining these traits—shell shape complexity of sutures, ornamentation, and hatching stage—Saunders and I finally had enough characters to perform a meaningful cladistic analysis of fossil nautiloids. To our surprise, these analyses showed the genus *Nautilus* to be extremely primitive. Rather than being a descendant of some Late Tertiary (perhaps 30-million-year-old) nautiloid genus such as *Eutrephoceras* or *Hercoglossa*, *Nautilus* appears to be derived from a much older ancestor, probably the Jurassic-aged, 180-million-year-old *Cenoceras* that our new genus *Allonautilus* so resembles. Furthermore, rather than being the last-evolved nautiloid genus, it seems to have been among the first of its family and the ancestor of most of the Cenozoic Era nautiluses. There was only one problem: If *Nautilus* was so old, why were there no fossils of it?

It turns out that there are many fossils that we can now confidently place in the genus *Nautilus*. The first to point this out was the paleontologist Richard Squires of California. Squires, a specialist in Tertiary-aged mollusks, was given a collection of nautiloid fossils from Washington state to analyze. These specimens belonged to a fossil species assigned to the genus *Eutrephoceras* by A. K. Miller in his famous 1947 treatise. Squires, however, was the first to declare that this particular emperor had no clothes. He had the intellectual courage to challenge Miller and the others by assigning his fossils to the genus *Nautilus*. Squires thus confirmed the 1957 discovery by the Russian Shimansky, as well as extending the known range of *Nautilus* back at least 40 million years. *Nautilus* now had a fossil record. But not enough of a one yet to jibe with the implications of our cladistic studies.

At about the same time that Squires's heresy was provoking the ghosts of A. K. Miller and Bernard Kummel to roll over in their graves, sheer chance played a part in this story. In 1988 I had the privilege of entertaining Steve Gould of Harvard University at my house for lunch prior to one of his speaking engagements. It was a perfect day, we had plenty of time, and because I had on my mantle a fossil nautiloid collected from Cretaceous deposits in California the year before, our conversation eventually turned to this particular specimen. So there it sat in my house, just waiting for Steve Gould to get invited to lunch. Steve looked at this large, beautifully preserved nautiloid fossil, and pronounced, "Looks like *Nautilus* to me." At the time I was aghast: Didn't the learned Steve Gould know there were no Cretaceous *Nautilus*? After all, he had worked alongside Bernie Kummel at Harvard for years and he was surely aware that there were not even any fossil *Nautilus*, let alone Cretaceous-aged specimens. But the comment stayed with me, and later, after Squires's paper about Tertiary-aged *Nautilus* from the west coast of North America was published, I began to wonder whether Gould and Squires were right. Might these west coast specimens be the missing fossil nautiloids that our cladisitic analyses suggested should be present? There was only one way to test this: Cut them open and examine their hatching stage, which would give us more characters with which to test their pedigree.

The Tertiary-aged species described by Squires and the Cretaceous-aged specimens from California, when cut, showed hatching stages identical to those of the modern nautilus. Like the modern nautilus, they hatched at

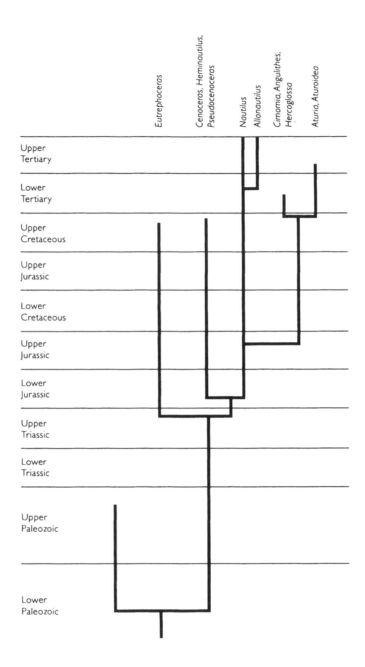

Evolutionary history of *Nautilus* and *Allonautilus*.

over an inch in diameter, with seven fully formed septa. Soon we found other nautiloid specimens squirreled away in various museum collections. Cut in half, they also showed internal shell characters that, when added to other shell and suture character states, gave telltale evidence of being—or in some cases not being—members of the genus *Nautilus*. We had shown that *Nautilus* lived in Cretaceous seas, and more recently we have found Jurassic specimens that indicate that this wondrous creature may have lived as along as 200 million years ago.

A final outcome of all of this is the need to suppress hundreds of species defined for various nautiloid genera. To date we can confirm only three (and perhaps four) discrete species of *Nautilus*: the modern-day *N. pompilius*, perhaps the modern day *N. macromphalus* (which may turn out to be a subspecies of *N. pompilius*), and a ribbed species from the Cretaceous of Vancouver Island. The others, including the hundred-million-year-old specimen that Steve Gould resurrected from my mantle (now sitting safely, if less eloquently, in a dark museum drawer) may all simply be *Nautilus pompilius*. On the basis of shell characters, that is certainly what they seem to be.

This raises a profound evolutionary question. Can any "species" exist for such a long period of time? Although there are many known cases of *genera* lasting for a hundred million years, the persistence of a *species* for so long is far more problematical, especially that of a species of a creature as complex as a cephalopod. Wouldn't genetic drift—the rather random evolutionary change that can affect any population—eventually result in a new species over such immense periods of time? Or would our hundred-million-year-old nautilus, if somehow brought back to life, happily and successfully mate with a *Nautilus pompilius* from today's oceans? Is a hundred-million-year-old species not a rare exception, but perhaps more common that we think? Are there even *billion*-year-old species on earth, perhaps among the bacteria, blue-green algas, or even flatworms, annelids, and mollusks as well?

9

Of Inoceramids
and Isotopes

To the north of Sucia Island lie the Canadian Gulf Islands, an isolated archipelago made up of the same Cretaceous strata we find on the U.S. side of the international boundary—the mixture of shale, sandstone, and conglomerate so typical of the Nanaimo Group formations. The Canadian Gulf Islands float like serene battleships off the coast, but getting to them is even harder than making port at Sucia. Unless you cheat (like the many smugglers in the region) and scoot across in a fast boat, the Gulf Islands are a long travel day from the United States. You have to take a British Columbia ferry out of the crowded Vancouver terminals, but once you are loaded aboard, conditions improve considerably; the sea voyage is spectacular as you head west, crossing the wide Strait of Georgia with the mountainous Vancouver Island looming in the distance. Once you arrive on these eagle-infested Canadian islands, it is clear that you are out of the United States: better beer, worse food, no Starbucks coffee (but mercifully no McDonalds either), and everywhere the lilt of Canadian English with the distinctive "ehs" creatively interspersed in every sentence. These islands are long and thin, carved by the glaciers with a north-south whim, and they form a rampart that seems to protect Vancouver Island from an encroaching North America. One of the longest of these islands is Gabriola Island. Its few, rare ammonite fossils tell us that its strata are not so old as the rocks of Sucia. Gabriola contains some of the youngest, and thus last-formed, rocks still of Mesozoic age in all of western Canada. Somewhere buried in its youngest rocks is a Cretaceous/Tertiary boundary site, the boundary between the Mesozoic and Cenozoic Eras now thought to have been formed by the impact of a giant comet 65 million years ago.

An ancient giant clam

The rocky northern coast of this island is largely uninhabited, although a road still gives access to it. One can easily find a trail through the salal and huckleberries, scrambling past the red-barked arbutus trees and onto a beach replete with strata. Black shale greats us, a shale dense with fossils and living oysters, and sometimes it is hard to tell the dead from the living. The fossils don't jump out at you, but they seem ubiquitous once you learn to spot them. The problem is that these fossils are almost unrecognizable as the remains of ancient life. They are large, usually the size of a dinner plate but sometimes as large as 3 feet across, yet so thin that they often appear as only slender white lines in the middle of stratal blocks. When we find a particularly large specimen that has been eroded into view, we can see that it looks like a gigantic oyster with a thin shell, the shell material covered with numerous fine ribs and some coarser undulations. Well-preserved specimens can be identified as some kind of giant clam. But how peculiar! There are no 3-foot clams on earth today with shells only 1/8 inch thick. These odd clams with nearly paper-thin shells are scattered in enormous profusion: the giant inoceramids, extinct for 65 million years.

Clams they are and huge as well, but they are not "giant clams" in the sense of the familiar giant clams, or *Tridacna*, so abundant in the tropics today. Tridacnas are large, *thick*-shelled creatures living among the coral reefs. The giant clams of these Canadian beaches are not even closely related to the living tridacnas. Instead, they were a kind of giant oyster—think of the chowder that could be made from such behemoths! The Gabriola Island beach we visit is absolutely packed with the things.

Gabriola Island is not unique in possessing a large number of gigantic clams of Cretaceous age. Sucia has many inoceramid clams (though none so large as those on Gabriola). In fact, most marine rock of Upper Cretaceous age anywhere in the world harbors inoceramid clams in varying, sometimes fantastic numbers. These clams may have been the most abundant type of larger life on earth in the Cretaceous Period.

Such mass occurrences are found elsewhere in the Vancouver Island region, such as on Hornby Island, Galiano Island, and some of the rivers far up the Strait of Georgia. They also occur in Texas, on a beautiful beach exposure in San Diego, and at many localities along the coast of Spain. It is always a spectacular sight: uncounted giant clam shells littering the ancient sediment. Sometimes these shells are themselves covered with smaller shells, the remains of encrusting clams and worms and other small creatures that used these islands of shell material as living quarters on the otherwise muddy bottoms. Nothing on the bottom of our world's oceans is even remotely comparable.

The central question is how these huge clams lived. Most, you see, are found in sedimentary deposits that appear to have accumulated in very deep water. In the oceans of today, there are very few clams of *any* sort in deep water, and certainly no clams up to 3 feet across. Here, then, is a mystery. The Cretaceous Period was a time when multitudes of giant clams lived in areas of the ocean where no self-respecting clam lives today. What sorts of creatures were these inoceramid clams? And perhaps more important, what kind of ocean did they live in?

An ancient giant clam the size of a plate. Cretaceous-aged inoceramid fossil, the most common type of bivalve shell from the Mesozoic Era, but now extinct.

We will need at least two types of time machines to answer these two questions. The first is called trophic group analysis, and it is a discipline of ecology—or paleoecology if we are analyzing ancient life. The second is a method that enables us to determine the temperatures of the ancient sea by analyzing fossil shells for their oxygen isotopic ratios.

The concept of community

We use trophic group analysis to examine how energy flows through ecosystems. All life on the planet depends on energy, be it from the sun, from chemicals such as methane, or from other organisms that are somehow ingested. Organisms eat one another, and as they do so, they pass energy from one body to another. You could say that the living associate along pathways of energy flow; such associations are often called communities.

Communities are generally defined as the recurrent associations of organisms through which energy flows. This flow of energy is quite often referred to as a food chain or a food web. Food chains and food webs are themselves broken down into groups called trophic levels, or feeding types. The lowest level of a community's "trophic structure" is composed of those organisms that use inorganic energy sources (such as sunshine or methane gas) to power the synthesis of inorganic compounds into living material through some chemical pathway. For example, plants carry on photosynthesis, transforming light energy and raw material into living material. Some types of bacteria use methane gas as an energy source. All such organisms are called autotrophs.

Autotrophs make up the base of the trophic structure in any community. They themselves are then consumed by plant (or bacterial) grazers, the herbivores. The energy stored in the living material of the autotroph's bodies is then transferred into new types of cells, those that make up the bodies of the herbivores. In turn, the herbivores are eaten by carnivores, and the smaller carnivores by even bigger carnivores; all, meanwhile, are being eaten by parasites. In this way energy flows throughout the system, and the organisms in this system are all considered members of the same community.

As we know, there are unequal numbers of organisms on earth, and these differences are often related to position in a food chain or food web. The most numerous, or voluminous (it is more practical to measure the total volume of a group of organisms than their individual numbers), are those organisms that capture and incorporate the primary energy source; they may be plants or bacteria. This volume is often referred to as biomass, and in most communities the biomass of plants is the largest in the system. Because the transfer of energy from plants to those animals that eat plants is never complete (most energy is lost), the biomass of herbivores is normally only a fraction of the autotroph biomass. In many ecological systems, herbivores account for only about 10% as much biomass as the autotroph. The same ratio obtains for the carnivores that eat the herbivores; they add up to perhaps 10% as much biomass as the herbivores.

If we arrange these biomass measures in tabular form, we quickly see that the trophic groups make up a pyramid. This trophic pyramid explains why large carnivores are so rare compared to herbivores and why plants are so abundant compared with any of the animals that eat them.

These and other simple principles are the rules that enable ecologists to organize their understanding of living organisms. They have been applied (with less success) to the ancient world as well, in the discipline of paleoecology. The main problem, as noted earlier, is that we can see only the animals that have left fossils, and thus we have a very biased view of the ancient world and its ecosystems.

These simple rules are the guiding principles of paleoecology. Just as particular mixtures of animals and plants in our world define living associations, or communities, so too should fossils found in recurring associations be considered ancient communities. The relative abundance of the fossils should similarly be clues to ancient trophic structure. And most important, the types of organisms should be clues to ancient environments if we accept the principle of uniformitarianism, which states that the present is a clue to the past. In our world, coral reefs are found in tropical latitudes in warm, clear, shallow seas. Coral reefs of the deep past *may* have occupied deep, cold water, but this is very unlikely, and much information suggests that reefs of the past inhabited environments very like those surrounding the reefs of today. In accordance with the principle of uniformitarianism, we use the lives of present reef animals to infer the biology of those now extinct.

The types of organisms found in any community, be it modern or ancient, are adapted to their specific environments. As environmental conditions change, so do the animals and plants that make up the various communities. All that said, what can we make of the inoceramid associations? Where in the food chain do they rest? What did they eat, and who ate them? Where do we find living analogs?

The mystery of the inoceramids

The ancient ecosystem of Sucia Island is fairly recognizable. Most of its fossils are made up of clams and snails that belonged to groups (if not to species) still living today. Most of the clams are forms that lived in the sediment, much like the vast majority of today's clams. These types of clams, which are called infaunal suspension feeders, burrow to escape predators. They feed by sucking large volumes of water through their necks and then straining planktonic organisms from this ingested water. Because most of the plankton is composed of single-celled plants, these types of clams are the herbivores of this ecosystem, the lowest rung of the food chain above the autotrophs, which in this case are the plankton. On Sucia, the clams are thus the lowest animal members of the trophic pyramid. Most of the snails from Sucia, on the other hand, were carnivores, and judging from the small bore-holes left in many of the ancient bivalves' shells, many of the snails fed on bivalves.

All in all, this assemblage of creatures on Sucia seems to make very good ecological sense. There are about ten times as many herbivorous forms (the clams and a few of the snails) as there are carnivores. There are a few oddballs—the inoceramids, for instance—that sat on the sediment rather than living in it. Yet these small forms (usually just a few inches long) are so much like oysters that they do not seem incongruous.

On Gabriola Island, however, and at so many other localities around the world with an abundance of *large* inoceramid clams, things are far more strange. The food chains as represented in the fossil record are not chains at all, for on Gabriola the giant clams are the *only* fossils. There were surely many soft-part organisms that left no fossil record, for there are many other signs of life, such as abundant trackways and trails, but other than the very rare ammonite fossils, it is a land of inoceramids. What is the nature of the trophic pyramid here? Perhaps the inoceramids fed on plankton and thus were the equivalent of the shallow-water clams. But there is a big problem with this scenario: The inoceramids appear to have lived at such great depths that they would have starved to death. There is no plankton pasture at the bottom of the sea.

A great deal of paleontological research has addressed this question, and the answer came from quite unexpected sources. The clues that helped us solve the mystery of the inoceramids came from two separate discoveries, one an act of genius, the other brought about completely by chance.

Taking the temperature of ancient seas

The act of genius occurred in Chicago in the early 1950s and was the creative work of a Nobel laureate from the University of Chicago, the chemist Harold Urey. Urey's discovery derived from the study of isotopes. An isotope is one of two or more atoms of a single element whose nuclei have the same number of protons but different numbers of neutrons. In an earlier chapter we saw how important the breakdown of radioactive isotopes is in geological age determination. Other isotopes have proved equally useful in a variety of geological investigations dealing with ancient environments. One of the most useful tools is oxygen isotopic ratios. Urey discovered that by measuring (with a mass spectrograph) the ratio of the very rare isotope oxygen-18 to the far more common oxygen-16 from calcareous shells, he could deduce the temperature at which the shell formed. The warmer the temperature of shell formation, the less oxygen-18 there was in the shells. As long as the shell has not been altered, the same types of analysis can be performed on fossil shells. Urey had devised a way to measure the temperatures of the ancient sea.

In the decades following this monumental discovery, thousands of scientific papers appeared detailing the results of oxygen isotope analyses of ancient temperature. Soon a rough, and then an increasingly more precise, picture emerged of global sea temperature through time. The most nearly continuous record of variation in marine temperature over the past 100 million years has come from isotopic analyses of well-preserved foraminiferans (single-celled protist shells) recovered from deep sea cores.

There were many caveats, of course. It did no good to compare the average bottom temperature (as measured from shells of bottom-dwelling creatures) and the average temperature of surface water (as measured from shells of planktonic creatures, which live mainly at the sea's surface), because bottom water is always much colder than that at the surface. But when similar sorts of comparative analyses were made, a quite detailed picture emerged.

The result of this work was to show that the temperature of the sea has clearly fluctuated. The two warmest periods of the last 100 million years were during the Late Cretaceous (about the time when the rocks that now make up Sucia were deposited) and during the 50-million-year-old Eocene Epoch. The world's oceans appear to have been appreciably warmer during these two times than they are now, even in high-latitude seas.

Many measurements of a variety of mollusk shells were made uneventfully, but when scientists began measuring the temperature of inoceramid shell material, they got a profound shock. The temperatures found in these shells were always very high compared to those of other shells in the same deposits. Inoceramid shells from Sucia Island, for example, yielded a temperature of 95°F! The only places on earth where one finds such hot sea water are right at the surface at the equator and in restricted lagoons and salt marshes. Sucia was tropical, though not that tropical, according to most indicators. Analysis of the other clams and shells from Sucia suggested that its sediments were deposited in fairly shallow water, perhaps 50 to 100 feet deep at most. Perhaps—just possibly—the water could have been that warm.

But a greater surprise was yet in store. Isotopic analyses of inoceramid shells collected from very deep-water sediment showed similarly high readings.

There was no possibility that these large inoceramids, such as the forms on Gabriola Island, lived in such shallow water. Unless . . .

Unless the giant clams did *not* live on the cold bottom at all but fell there only after death. In the 1980s a new hypothesis was proposed: Inoceramid bivalves were attached to floating logs, and all those found in various deep-water deposits had arrived there only after they died. There seemed to be no other way to explain the isotopic results, and many serious scientists subscribed to the "floating inoceramus" theory.

In the late 1980s, two of my graduate students advanced another idea. In a paper published in the journal *Geology,* Ken MacLeod and Kathryn Hoppe, after analyzing inoceramid shells collected from deep-water cores in Europe, suggested that the inoceramids formed their shells in a way that excluded the isotope oxygen-18. This led to spurious results when the values were converted to temperatures using the methods of Urey. The inoceramids were *not* living in 95°F water, but their shells were painting that picture through what was described as a "vital effect."

Such vital effects were not unprecedented. They are also known from living corals, which harbor a tiny plant in their flesh that aids in formation of the skeleton. The presence of this microscopic alga in the coral's flesh also reduces the amount of oxygen-18 incorporated into the coral's skeleton, and this yields an anomalously high paleotemperature reading. Yet corals live only in very shallow seas, for the symbiotic algas in their flesh need warm water and light. The deep-water inoceramids could not possibly be harboring such shallow-living symbionts. Something else had to be involved.

Hydrothermal vents and "cold seeps"

The answer came from a completely chance discovery: that of hydrothermal vent faunas in the modern-day deep sea. On February 17, 1977, the submersible ALVIN was diving on a mid-ocean ridge deep in the Pacific Ocean. These are the regions where new sea floor is produced. Oceanographer Tjeerd van Andel, who was on this historic dive, described the discovery as follows:

Whereas the ocean floor at this depth [2500 meters] tends to be rather poor in organisms large enough to be seen from the ALVIN, the spring [a hot, volcanic water seep] seemed to be a veritable garden or aquarium full of large life forms. White clams, mussels, large white crabs resembling those of coastal tide pools, worms with featherdusters living in calcareous tubes were everywhere. The number of animals, the total amount or biomass, seemed far, far larger than anything ever seen on the deep sea floor. Moreover, most of the animals did not resemble the deep sea kind—sea cucumbers, anemones, shrimp—at all, but rather reminded me of the fauna of a coastal tide pool. On the first dive I was especially struck by the white clams with enormous shells measuring up to 20 cm in length; we later found them to weigh several pounds each. (Van Andel, 1977, pp. 149–150)

All of this newly discovered "vent fauna" lived well below the zone of plankton, just like the inoceramids of the Cretaceous. The vent fauna, and faunas discovered soon after in regions of cold methane seeps (areas where cold, methane-rich water comes up through the sea floor), used an entirely different energy source than sunlight for the base of their food chains. They used methane gas.

The discovery of deep-sea vent faunas revolutionized not only biological oceanography but in a significant way paleontology as well. Hydrothermal vent and cold seep-faunas (where animals cluster around natural gas, cold brine, or petroleum seeps) are diverse and rich in life in places where life is normally sparse and rare.

The discovery of hydrothermal vent communities in the late 1970s alerted biologists to the presence of significant associations of "chemoautotrophic" organisms on the sea floor—organisms that use as their primary energy source not sunlight but other types of energy. By the 1980s it was realized that two distinct associations of animals are present: hydrothermal vent faunas and "cold-seep" faunas. Cold seeps are found in a variety of volcanic areas and over organic-rich sediment accumulations. Cold seeps produce a very peculiar type of sediment formation; they produce limestone bodies in sediment where limestone is usually absent. They also contain a fauna different from that in the surrounding sediment.

Present-day hydrothermal vents and cold seeps are characterized by localized concentrations of hydrogen sulfide and/or methane-rich fluid that are generated and then expelled onto the sea floor. At these sites, high densities of chemoautotrophic bacteria that live on methane occur in suspended vent-fluid emissions, or as bacterial symbionts enclosed within invertebrate tissues, or even as free-living surface mats. The chemoautotrophic bacteria oxidize the vent-seep fluids to produce metabolites and energy at the base of a chemosynthetic food chain. These vent fluids mix with sea water to form localized and often anomalous sedimentary precipitates that include metallic sulfides and barite, and isotopically distinctive carbonates. Many of the organisms adapted to these areas live in environmental conditions that would prove toxic to most marine organisms.

The recognition of vent and cold-seep localities in the ocean floor caused paleontologists to re-evaluate the origin of numerous anomalous rock and faunal associations. Most troubling had been the presence of isolated limestone bodies in deep-water sandstone or turbidites. These had long been interpreted as shallow shoals or fossiliferous blocks transported to deep-water settings. The fossil fauna within these limestone bodies was quite different from faunal elements found in the surrounding matrix, just as it is in modern cold-seep areas. The final peculiar characteristic of these limestone bodies was their isotopic composition. They yielded isotopic readings typical of very warm water, when in fact they formed in water near the freezing point. Limestones in modern seep deposits, just like the shells of the ancient inoceramids, give a false picture of their temperature of formation.

The fauna of such cold seeps are now known to be highly distinctive in both their recent incarnations and the distant past. Most common and prominent members of extant cold-seep communities are large clams and mussels with bacteria in their flesh that enable the clams to live off methane in the surrounding sea water rather than off plankton, like most other clams.

Tube worms and specialized gastropods are present as well. One of the most striking aspects of the cold-seep bivalves is their large size. The bivalves found in the seeps are normally several times larger than any bivalves found in the surrounding, non-seep deposit sediment.

Since these early discoveries, a variety of cold-seep faunas, mainly of Mesozoic and Cenozoic vintage, have been identified in the geological record. Yet research into modern and ancient cold-seep accumulations is still in its infancy. To date, research efforts have been largely observational; we have little or no theory from which to predict the geographic placement, biological makeup, or temporal duration of these deposits.

A large number of cold-seep regions have now been identified. Many are associated with deep-sea plate tectonic settings, although they also occur on the margins of continents, at the base of underwater cliffs, and at oil and gas seeps on the continental slope.

The fauna surrounding vents and seeps is normally characterized by an unusual taxonomic assemblage. Some researchers have proposed that vent/seep faunas are long-lived groups composed of many relict taxa of great geological antiquity. This implies that vent/seep faunas have been largely immune to the mass extinction events that have so profoundly affected the rest of our planet's biota. If so, the vent/seep fauna may be the most insulated group of animals on earth from mass extinction events, being largely self-sufficient except for a need for oxygen. This fauna might be able to survive even catastrophic meteor impacts and other planetary calamities. Vent and seeps may thus be important faunal reserves or refuges, "lost worlds" retaining archaic forms. An important and still unresolved question is whether the vent/seep areas are also seed stocks for evolutionary innovation and planetary repopulation following a mass extinction event. Are these faunas units of evolutionary innovation, where new types of life first arise? Are they regions of relict taxa? Are they sites impervious to mass extinction, and therefore storehouses of diversity?

Fossil cold-seep areas are most commonly recognized from Tertiary deposits. Far fewer have been recognized from Mesozoic deposits, and only small number are yet recognized from Paleozoic deposits. Verena Tunnicliffe of the University of Victoria, a specialist on this fauna, has proposed that the vent/seep communities first arose in the middle Paleozoic, although this theory has never been rigorously tested. Cold-seep deposits dating from the early Paleozoic and older are still problematic.

The inoceramid clam accumulates are surely related to cold seeps. I surmise that they were clams that housed symbiotic algas or (more probably) bacteria and lived preferentially in areas of the sea bottom where methane gas was being released. The fact that so many regions of the Cretaceous oceans seem to have been colonized by inoceramid clams, presumably all living on methane, tells us something fascinating about that long ago time. The oceans then were clearly different in chemical structure. They were far less homogeneous, they were more turbid, and they retained organic material in sediment to a far greater degree than now.

The pristine inoceramid beaches of Gabriola Island have given us a priceless glimpse into the deep past—thanks to an isotopic time machine or two.

Part Four

The Time Machine

10

Cretaceous Park

We have come a long way, in space and time, through the use of the various time machines profiled in this book. Fossil, radiometric and magnetic dating machines; the detection of ancient latitude, sea level, and temperature; the reconstruction of fossils and the communities they lived in. Yet there are so many more invaluable techniques that we have not even touched up here.

How to summarize? Perhaps the best way to recap the discoveries, reinterpretations, and even speculations that this book recounts is to recreate ancient Sucia Island in a story, using the disparate facts we have unearthed as the backdrop. To do that I'll use one of science fiction's favorite devices, an "actual" time machine.

Physicists, the top carnivores of the trophic pyramid of scientists, have at their service the most expensive machines and instruments ever devised. To smash atoms, to find and identify the smallest particles in the universe, to see and measure the last whispers emanating from the Big Bang—such quests require brute instrumental strength. Nevertheless, physicists still routinely encounter phenomena and questions that are beyond the scope of even the most advanced technology. When this happens they must take an entirely different tack—an approach that may not solve a problem but that at least brings it to an intellectual point where it can be re-examined and perhaps can eventually be solved theoretically rather than experimentally or observationally. These exercises are "thought experiments." Einstein loved thought experiments—and needed them. Yet these devices are not the sole property of physicists. Scientists in other disciplines (such as paleontology) can use them too. Every time a paleontologist attempts to reconstruct conditions in some long-ago time, he or she is conducting a thought experiment.

Our concluding chapter is science fiction, of course, and so is sure to incite the wrath of those who cringe at any bending of the unspoken rules governing content in "trade science books" (books written, in other than a textbook format, for people who love science). But, as the preceding chapters will bear witness, the *science* in the following science fiction comes from many years of dogged detective work by many different scientists. What follows *is* a thought experiment of sorts, for it is the best guess I can make, on the basis of the evidence, about what a true time-travel expedition back to the locality of Sucia Island in the Late Mesozoic world would encounter.

June 15, 2222

The device sat gleaming under sterile fluorescent lights. It looked rather like a boat, instead of whatever a time machine is supposed to look like. And that was a good thing, actually, given that its prime purpose *after* time-traveling to the Late Cretaceous Pacific Ocean, just off the western coast of North America, was to float reliably for three hours in a long-ago sea.

The event itself was easy: One moment the operator was in his own time, sitting in the cockpit with all of his gear piled around him, and the next he was bobbing on the surface of an ancient sea. His first look was instinctive: The numbers on the chronometer—the record of how far back into time he had traveled—stared back at him. His own time was such a low number for such an old planet—to measure time in the number of years since the birth of a man seems ignorant of how vanishingly brief has been our human sojourn on earth, but a system of time measurement built on historical precedent is an intractable master. Hence the date on the console read $-76,600,000 \pm 100,000$: seventy-six million years before his time. To the right of these numbers were the other displays, based on other types of time keeping. The first display was marked Geomagnetic Polarity Time Scale: It read magnetochron 33R, which told the operator that he was in the first magnetic reversal after the Cretaceous long-normal interval. The display marked Biostratigraphy Time Scale was subdivided into European Standard, Western Interior, and Pacific Coast columns and was designed to indicate the names of the stages and fossil zones. They read Campanian Stage: *Bostrychoceras polyplocum* Zone, *Baculites scotti* Zone, *and Baculites inornatus* Zone, the zones being named after the diagnostic fossil ammonites found in Europe, the interior of North America, and the Pacific Coast of North America, respectively. At the far right, the instrument panel listed the sea level indicator: Global sea level sequence Highstand 22. He stared at this panoply of measurements of time, as recorded by years, magnetic reversal stratigraphy, fossils, and the very level of the sea itself in its global basins, trying to decide whether he felt any older, having become the first human on earth. Yet the mishmash of numbers and letters only elicited a wry chuckle, for he, now Master of all Time, was still its slave: Enmeshed in time's various reversible threads was the most inexorable timekeeper of all, the wholly finite and utterly determinate number of heartbeats measuring his own life—a stream that could never to be reversed.

The repetitive commands of long training took over, and he found himself pulling out the inflatable buoy with its line and anchor. He watched himself, as if from a remote distance, setting the buoy with its bright strobing beacon. The anchor snagged quickly, telling him that he was in shallow water, just as the planners had promised. His insertion point (and he certainly hoped, his removal point as well) was now marked in this sea, and at last he could look around on this fine afternoon, late in the Cretaceous Period.

First impressions: Land was visible nearby, perhaps a hundred yards off; it was green and lush, with tall trees and a thicker understory. The odor of fetid jungle and other things unknown to him thrust through the scent of the sea; the very air was a miasma of pines, flowers, swamp, methane, saurians, and a thousand other Mesozoic perfumes no man before had ever smelled. He breathed deeply of this Cretaceous air, both attracted and repelled, and felt—or imagined he felt—immensely invigorated. He pulled out a small bottle, opened it to the air around him, and then sealed it, a first sample. He looked once more toward land and saw a faint surf painting white streaks on the distant beach. Farther inland stood high mountains, and one of them trailed a thin plume of black smoke and ash, the atmospheric signature of an active volcano.

Yet these were but the briefest of impressions before a new sight captured his attention. Movement caught his eye, and he stared up into a sky full of wonder—winged wonder—for above him wheeled great apparitions of scale and fur, that could only be pterosaurs. These bat-like creatures, somehow simultaneously monstrous and benign, were flying—not gliding—majestically overhead, with many smaller reptilian and avian fliers moving more rapidly among them. He was amazed at the size of the larger reptilian aviators; it was as though small airplanes had become sentient and now navigated the sky unfettered by human constraint. His small boat bobbed in the swell, and our time traveler marveled as the great and small fliers skimmed the surface for fish, or dived headlong into the sea, like great pelicans at play in a hallucinogenic airshow.

"It would be good just to float here," he thought, "offshore of this Mesozoic land, and simply watch the aerial circus above." But he had been a diver for many years, and always the internal clock was ticking, whispering in his ear about time passing; this voyage nearly 80 million years into the past was like a scuba dive: borrowed time in a foreign world. He had work to do.

The first observation was the quickest, although he found that taking a latitudinal measurement with a sextant in a rocking boat was trickier than he had anticipated. Nevertheless, he shot a crude solar. He didn't need great precision, for the number he recorded only confirmed what his first view of the sun had already told him: He was at a low latitude, surely the tropics, somewhere between 25 and 30 degrees north. Nearly 80 million years hence, this place, which would be named Sucia Island by an eighteenth-century Spanish explorer, would be marked on the first charts of the region at 49 degrees north. Between the end of the Cretaceous and the emergence of *Homo sapiens* some 65 million years later, Sucia Island and all of its surrounding territory would travel more than 2000 miles north!

He glanced at his checklist and began preparing for the dive. The twin-80 tanks felt marvelously light in their titanium casings, and he decided to wear the thinnest wet suit after measuring the surface water temperature and finding it to be a balmy 89°F, compared with readings of less than half that in his time. He looked around once more, tasted twenty-third century air as he sucked from his regulator, and rolled into the sea.

An ancient sea bottom

A lurch and a loud splash, the usual momentary vertigo while he somersaulted, and then he bobbed toward the surface. The water was turbid, filled with great clouds of plankton. He was anxious to get down below this murky surface, for in *his* world the surface regions were patroled by large carnivorous hunters, such as the sharks, and he had no doubt that this world's larger carnivores, which included mosasaurs and elasmosaurs as well as sharks, were far more dangerous than anything of his time. He made a final check of his gear, zeroed his watch, traded snorkel for regulator, and slid down into the late Mesozoic ocean.

Underneath the sea's turbid surface layer visibility improved, and our time traveler was relieved to see the dim outlines of the bottom, far below. He had never liked vertical descents, especially in very deep or murky water where you lost sight of both the surface and the bottom. He swam down quickly, chased by beams of shimmering sunlight, through a cerulean-blue water column rich in tiny invertebrates and minnows, and he thought of the hundred new species he was surely passing by. Now the bottom was visible, coming up fast, and he gently dropped down onto its sandy surface, landing on his knees, to be surrounded like a dream by a Mesozoic sea bottom festooned with multicolored algas: bright, current-wafted pennants in browns, yellows, reds, and greens. But his eyes ignored the submarine flora; he searched instead for old friends he had known only in death. He quickly saw them in legions, quite alive, and he greeted them with the hoots of joy that are the only underwater exclamation of a scuba diver.

The sandy bottom was slightly rippled, barely disturbed by the stronger waves at the surface. The tubular, tentacled necks of clams whose shells were deeply buried in the fine sand were scattered across the bottom in great number, and he wryly wondered what a chowder of these Cretaceous clams would taste like. Snails large and small crawled by, leaving irregular tracks across the sediment surface, while circular sea urchins rattled their pike-like spines among the immobile inhabitants: the sponges, bryozoans, and hydrozoans and a myriad of other colonial and solitary filter-feeding animals living in their skeletal mansions and tenements. But of all the creatures visible, one stood out. The bottom was littered with large clams both dead and alive. These Cretaceous clams were distant relatives of the oyster and perhaps the most ubiquitous residents of the Late Cretaceous seaways. By the end of the Cretaceous Period, 65 million years before the time of humans, they would be entirely extinct. But on this older sea bottom they were everywhere. They sat atop the sediment, not buried in it like the other species of clams here, and their large, ribbed shells were crowded with colonies of encrusting invertebrates. Bright mantle flesh extended partly over the inoceramid's shells, making them look like the giant clam *Tridacna* of modern coral reefs, and probably for the same reason, he thought: *Tridacna* clams harbor gardens of symbiotic, single-celled phytoplankton in their flesh, which aid in respiration and shell formation. Many scientists thought that the inoceramids used the same trick, or perhaps they used bacteria instead. "We shall see," he told himself.

He took a small sample of the sediment and then placed an entire inoceramid, shell and all, in the large pouch sewn into his buoyancy compensator. With these samples packed away, he shoved off the bottom, taking photos with his camera as he rose, thinking of the papers that would be written about these clams and thinking of other things: his air supply, the depth, the time, and the other animals he needed to find.

The dive was a great mix of the unknown and the familiar. The sandy bottom was like many he had seen before, in his own time. The fields of clam necks, the tropical mollusks and echinoderms—much of what he saw had a perfectly modern appearance. But other things, such as the inoceramids and the partly buried trigoniid clams, were foreign—anomalies that were jarring to his trained eye. They were so out of context. They should be fossils in an outcrop, not loosely scattered on this bottom, not so obviously alive.

He powered on, his large fins and steady kick carrying him rapidly over fertile fields of the clam-rich community. He decided to descend further and began to follow the sloping bottom into deeper water, looking for those creatures he had so long dreamed about and so long studied, the Mesozoic swimming shellfish called ammonites.

He knew what they ought to look like. Some had shells like great wagon wheels with the body of a squid stuffed inside ("*Perhaps!*" he told himself) or, more probably, like that of the chambered nautilus of his own time, the last remaining member of the long-lived group of now largely extinct cephalopods to which the ammonites belonged. Others had evolved more ambiguous shapes—some straight, some like snails, some like candy canes, some like the mess a child makes with a capless tube of toothpaste—and no scientist knew what these uncoiled forms did for a living or even how or where in the ocean they lived. About the mysterious ammonites only one thing was certain: All of them, irrespective of shape, died out in the same great catastrophe that killed the dinosaurs, a calamity brought about by the dire environmental effects immediately following the collision of a large comet with our planet, some 65 million years ago—a calamity that set the stage for a new suite of creatures on land and sea to vie for dominance.

But here on this quiet bottom, with the sunlight from the sea's surface 50 feet overhead mottling the sandy bottom and its late Mesozoic inhabitants with faintly visible patterns, that catastrophe was still 12 million years in the future, and ammonites and their Mesozoic world were in fine fettle. He had already seen in the sand shell fragments that he suspected were from ammonites, and if he couldn't find any living specimens, these would have to do. There were too many questions waiting to be answered for him not to bring home something of these enigmatic animals, even if it was only fragmentary shell material from long-dead specimens. But the living could answer far more questions than the dead, even the recently dead, so he swam onward, searching.

Ahead, emerging from the gloom, he saw a new feature: what appeared to be a large mound of rock. Yet, as he moved closer, he saw that the "rock" moved ever so slightly in the swell, belying its lithic appearance. With a start he realized that he was looking at a carcass of rotting flesh and disarticulating bones, a once-large marine reptile—an elasmosaur?—now transformed into a submarine feeding station. It was the focal point for a diversity of scavengers and carrion eaters. Sharks swam among the bones, worrying slabs of flesh off the rib cage, while smaller fish nipped at gray carrion. Some areas of the carcass were nearly covered with crabs and snails, scavengers cloaked in a carnival of shape and color. The sharks looked quite familiar, as did the crabs, but many of the snails were strange to him. These ornate creatures had ribs and knobs like those found today only among tropical mollusks, but they were members of species no longer alive.

He moved closer, wary of the sharks (which looked just like those of his world), and as he did so, a new side of the carcass came into view. Upon seeing the new suite of animals feeding there, he felt a great throb of joy, for among the busy scavengers were several old friends, creatures still alive in the modern-day tropical Pacific: pearly nautiluses. Two large adults and several juveniles belonging to at least two species chewed on a large hunk of flesh, and he could hear the sound of their heavy beaks rasping on a bone's edge.

He had expected to see nautiloids, but he had not expected them to seem identical to the nautiloids of his world. Somewhere in time, he thought, Eldredge and Gould must be smiling—and Darwin rolling over in his grave. These species would not change in any appreciable fashion for more than 80 million years. They would survive so much, including the impact of the comet that spelled the dinosaurs' demise. They would become "living fossils," persevering unchanged. But this species of the genus *Nautilus* would also give rise to a whole slew of new species in the aftermath of the impact event that ended the Age of Dinosaurs.

He moved in closer to the oblivious nautiloids and pried one away one from its meal. In color and shape it seemed identical to *Nautilus pompilius*, the familiar Nautilus found in the Philippines, Fiji, Australia, New Guinea, and Samoa and on a thousand smaller islands and rocky reefs across a huge expanse of the Indo-Pacific. He let it go, and it turned back toward its meal, tentacles extending outward. His tank was now nearly half empty, and he had not glimpsed the quarry he had traveled so far to see. Were the ammonites restricted to far deeper water than he could visit? Their fossils were common enough in the sedimentary strata of the future Sucia Island deposits, yet here, on this bottom that was to become those sedimentary rocks, he saw few of their shells, and he saw none alive.

He again followed the sloping sea floor into greater depths, and as he did so, the creatures on the bottom changed. As the sediment became finer, the number of burrowing clams greatly diminished, as did the numbers of snails and other creatures that he had seen living atop the sediment. The fine mud at this 100-foot depth was streaked by a diversity of markings and trails left by vermiferan life forms; here and there a slow-moving sea cucumber or toiling crustaceans also could be seen. The only shelled creatures were the ubiquitous flat clams, the inoceramids, but even they were reduced in number. In the distance he saw another nautilus, and he marveled anew at how much they looked like others of their kind that he had trapped in the

Philippines, New Guinea and Fiji and had dived with at night in the cool water outside the great New Caledonian barrier reef, the only place on earth in the Age of Humanity where nautiluses could be seen in the depth ranges of a scuba diver. But in those places, the nautilus lived on the deep fore-reef slopes, the muddy depths in front of coral reefs. Here they inhabited a completely different environment. He had not seen any coral at all, and despite the warmth of the seas, there were no reefs. Another mystery among many.

In spite of the bright sun he knew to be shining far above, he was now gliding through a twilight world, a dim smoky topaz of clarity but little light. Yet it was by no means silent: The snapping and crackling of crustaceans in thousands of burrows around him produced a chaotic percussive symphony. This was his only company as he glided ever deeper, and he felt very, very alone—the only human on the planet, the only mammal larger than a squirrel.

The water became clearer and colder as he passed through a thermocline. Now he could see for many tens of yards in all directions. Ahead of him, the greater depths of the slope he descended yawned like a dark night beckoning. A large shape looming to his right startled him, and it resolved into a slim shark following its own agenda, passing through the sea whips and sea fans encrusting the rare rocks and larger shells that offered anchorage amid the pocked and burrowed mud. A school of small lobe-finned coelacanth fishes fled before the shark and then resumed their own hunting.

An old friend

He reached 150 feet, his maximum depth limit on this dive, and now his air supply and his body's nitrogen uptake became primary concerns. But those worries vanished instantly as a small forest of objects resembling sticks and large pencils descended to the sea floor about 10 yards to his right. He was mystified; there was nothing in his long diving experience to compare with the sight. He checked his camera settings, turned the camera strobe to maximum power, and glided toward this curious apparition, uncertain what these creatures were, but unafraid. He approached the nearest of the sticks and saw that it was not wood but rather was composed of a long, tapering cone-like shell, with many thin tentacles extending outward from the oval opening at its bottom. The tentacles attached to a tubercular head adorned with fleshy ridges and two large, unblinking eyes recalling H. G. Wells's vision of Martians. But these were not Martians. They were earthlings of long standing that first evolved in the Devonian Period, some 400 million years before humanity. They were ammonites, in this case straight-shelled forms, the ubiquitous *Baculites inornatus*, by far the most common (and diverse) ammonites of the late Cretaceous seas.

He glided in close to the nearest baculite. Its shell was about a foot long, and the creature and shell were oriented in a nearly vertical position in the sea, perpendicular to the bottom, with the head and tentacles facing the bottom while the sharply tipped apex of the straight shell pointed toward the surface far above. Here was a creature that lived an entirely vertical existence, rising or falling in the water column but having little ability to swim laterally. Its very design suggested speed, but speed upward, and he guessed that these ammonites responded to threats by quickly taking flight in the unexpected vertical direction.

He knew this particular species well. It was an old acquaintance that he had found as fossils in so many places of his world: the Sacramento and San Joaquin Valleys of California, Baja California, Colorado, South Dakota, and even farther afield, in Madagascar, South Africa, and Chile. Here it was one of a large school, busily rooting through the soft sediment with its tentacles. Occasionally it would find some crustacean morsel, and pass it to a mouth located in the center of the tentacles.

Our time traveler thought it one of the most beautiful animals he had ever seen. It was not like a Nautilus at all; it gave none of the impression of stupidity and lassitude that set the nautiloids apart from the rest of the cephalopod clan. It was much more akin to the most advanced and beautiful squid imaginable, and the streamlined cone of its shell only enhanced this sense of modernity, not antiquity. He had seen a thousand of its fossil shells, but somehow the living creature was a revelation, like the first view of Saturn or Jupiter through a good telescope. Pictures—and fossils—simply did not do justice.

He watched a play of chromatophores on the baculites's head region—shivers of color passing from tentacles to head, increasing in strength and hue when a particularly large crustacean was found. He came closer, and the creature remained oblivious, or indifferent, to his presence. He was now only inches away, lying prone on the soft sediment, watching this ancient organism feed, one of hundreds similarly engaged on this muddy bottom. He tried to peer into the mantle cavity and felt almost like a voyeur, but he had to know. "How many gills? Where's the ink gland? What does the septal mantle look like? How does it make its intricate and florid sutures, among the most complicated structures ever produced by nature? And what are the sutures for?" He swung the Nikonos underwater camera with its flash unit around, centered the nearest baculite in his viewfinder, and shot pointblank at the long, thin ammonite.

The flash, a detonation of light in this near-darkness, precipitated the most amazing spectacle. A hundred slim cones blasted off the bottom as one, the entire school streaking up in concert, leaving a great trail of black ink in their wake, and thus looking for all the world like an ICBM strike in progress. He watched them rise to 20 or 30 feet above the bottom. There they slowed, finally hung motionless in the water column, and then began sinking slowly back down to the bottom, still vertical, a hundred rockets forgetting their fright, returning to their launching pads—and feeding grounds. He savored the sight and shivered with pleasure, or was it the invading cold?

The cold was a wake-up call; it was clearly time to go. He rose up with his bubbles, carefully following the smallest and slowest, ascending into warmer and brighter water, and soon neither the bottom nor the surface was visible. He ascended in this cocoon, watching jellyfish, small herring, and once a compressed, planispiral ammonite swimming just at the edge of visibility.

Near the 30-foot level he began hearing sharp cracking noises, and he pirouetted to scan the water around him. He was surprised to see powdery and chunky white and brown material falling around him. A larger piece cartwheeled downward several yards away, and he interrupted his ascent to swim after the falling material, catching it at a depth of about 40 feet. It was the size of a dinner plate, with irregular edges. One side of the fragment gleamed silvery in the shafts of muted sunlight dancing down from the surface, and he saw that its other side was creamy white, with a large reddish stripe traversing one corner. There was no mistaking this shell. Somewhere above him, a large ammonite had just died.

He dropped the piece of shell and warily rose to 20 feet to start his decompression stop, slowly turning 360's in lookout. During his second or third rotation he saw a distant school of creatures coming in his direction. As they approached through the shallow surface turbidity, he recognized them as planispiral ammonites, species with large discoidal shells, all swimming backwards with head behind. Some had shells more than a yard in diameter, and all were countershaded: dark color on top of the shells, white beneath, a camouflage evolved because it fools predators either above or below. Pachydiscids, he thought, seeing the distinctive ribbed shells. They swam slowly but purposefully, conveying a sense of power, like old battleships passing in stately formation. They were not like the skittish, torpedo squids of his day, which relied on speed to save their unprotected and succulent bodies from becoming meals. The ammonites used their shells like ancient dreadnoughts, a streamlined but massive armor. He could now see their large, octopus-like eyes as they passed by, the closer animals veering to avoid him, the farther reaches of the school paying him no heed.

He watched the last of the large ammonites swim lugubriously into the distance, finally to disappear, swallowed by the sea as they left his 50-foot circle of visibility. He was nearing the end of his first decompression stop, and as he watched the second hand on his chronometer sweep its slow arcs, movement caught his eye. He saw another school of large ammonites just at the limit of visibility, and once again he heard a piercing crack. He began to swim toward this second school, the bright sunlit water warming his body.

The ammonites were careening and scattering as he approached, and now he could see a larger, wraith-like shape among them. He backpedaled as he watched the mosasaur feeding among the ammonites. It was relatively small, only about 8 to 10 feet long, and its shape and head were quite familiar to him. He was well acquainted with the monitor lizards of Pacific islands in his time (a 3-footer had once crawled into his bed on an island on Australia's Great Barrier Reef), and he had even seen a Komodo dragon, the largest lizard of his world and found only on a remote island in Indonesia. But even the Komodo dragon, which could reach a length of 8 feet or more, was puny compared to the more massively built and far longer mosasaurs, which could reach a length of 30 feet. Yet the mosasaurs were close relatives—direct ancestors, really—of those future lizards, and although of dinosaurian dimensions in some cases, they were simply lizards grown giant. But their feet had evolved into paddles, and their tails had broadened into a large, flattened shape more resembling that of a crocodiles than the long whip-like tails of its terrestrial relatives. He watched it swim, admiring its speed and agility; in its element, the mosasaur reminded him of the large seagoing crocodiles he had seen in Micronesia, creatures far more dangerous to humans than any shark ever evolved.

Although he thought he could fend off this particular mosasaur if need be, he didn't want to take the chance. But the marine lizard seemed unconcerned with him, if it had seen him at all; it dashed again into the crowd of far slower ammonites, taking the body chamber of one of the larger ammonite's shells in its jaws. With a rending crack the shell broke, and the slashing jaws quickly tore into the now-exposed body portions of the ammonite. He watched the victim disappear in an explosion of blue blood and then slide down the gullet of the lizard as it turned toward another fleeing ammonite.

He swam quickly forward and down and snagged the largest piece of falling shell. He looked for tooth marks or any evidence that would alert some future paleontologist to the true nature of the predatory attack he had just witnessed. Like the first bit of shell he had caught earlier in his dive, this fragment bore no holes.

The ammonites retreated into the distance, still menaced by the small mosasaur, and once again he was alone in the uppermost region of the Cretaceous ocean. A brisk wind had risen during his dive, and he was buffeted by the wave action only 10 feet above his head. He looked again at his watch and saw that he was now free to surface. He watched small fish skim just beneath the wavetops and found himself trying to see the smaller life forms around him, wishing he had some way to look at the plankton and search for the juveniles or larvas of creatures whose adult forms were familiar to him. He knew that tiny ammonites, clams, snails, urchins, crabs, and at least a thousand other species surrounded him in these highest portion of the sea, but he could see only particulate matter floating by. Out of excuses now, and nearly out of air, he surfaced, seeing once again the blue-green sky, the scudding clouds, and the diverse and spectacular reptilian and avian pilots soaring over his head.

He was well out to sea from his boat, which was just visible in the distance, a winking reminder of his other life, his real life. He began to swim toward it and looked up at the sun. It was far lower in the sky and was descending toward the ocean to his . . . west? He stopped, pulled out his compass, and laughed through his snorkel. According to his instrument, the sun, now plummeting toward the horizon, was sinking into the east. "First time a human has seen that," he thought to himself, remembering that this was a time of reversed magnetic polarity. North was south here, and east was west.

When he finally reached the tethered boat, he was exhausted from the long dive and the longer swim. He released straps and slid out of his tank harness, free now and unfettered. After resting several minutes, just for good measure he lay prone, hyperventilated, and piked at the waist. He slid downward once again, free-diving without a tank, the first marine mammal in this world, and cranked off a 30-foot dive in spite of his fatigue. He got a last glimpse of the Cretaceous sea bottom and was rewarded with the sight of a last ammonite, a heavily ornamented *Hoplitoplacenticeras*. Coming up, he again passed through the zone of plankton, and he paused only a meter below the surface to look more closely at the thick plankton soup. Near his face mask were unnumbered small round shells, a millimeter or two in diameter, with tiny tentacle faces extending out of the apertures. With a shock of recognition, he realized he was in a sea of newly hatched ammonites, thousands living in the top few meters of the sea at this early stage of their lives. Lungs burning, he returned to the surface.

He had only to wait now, and this journey would be over. But he looked toward the land and thought about the inland sea that must exist just behind those coastal mountains, a sea that would mimic the much later Sea of Cortez, which he knew to be filled with giant reefs of clams, not corals, and with ammonites from Texas, such as the great snail-like *Didymoceras* and *Nostoceras*, and with the largest ammonites of western North America, *Pachydicus catarinae*, named for a tiny fishing village in Baja California. He knew the future Rosario formation of Baja lay only a short distance to the east ("but west by my compass," he laughed to himself)—inland, anyway, into the heart of the North American continent—and he thought he could get there easily and see dinosaurs to boot, realizing a final dream. But who would ever know about these wonders? He could never return to his own time if he took that voyage. He would learn much, even if he lived only another day here. But was it worth it to know the answer to a question if you could never tell any other human about it? Was science simply learning the answer? Or was dissemination of the results a vital part as well?

As he settled into the cockpit, he reflected on such matters and on everything he had seen during his time travels. The informed imagination, he decided, may be the best time machine of all.

Afterword

Given full throttle the engine roared, and we skimmed over the glassy flat sea. I watched the cliffs of Sucia Island fall away. The island's dark-shale coastline rapidly receded, changing from a pile of acutely visible sedimentary rock to a more ambiguous low shape.

The boat was crowded: Tom Daniel and his family were aboard, as were my wife and new baby boy. We had just spent two wonderful days on Sucia, camping on its northern shore, finding fossils during the day, and watching the bright stars each night before retiring. Now, the weekend finished, we headed east toward our launching site.

Other islands began to snake by now, smaller bits and pieces of Wrangellia. But somewhere during this hour-long voyage we crossed a great boundary, the fault marking the edge of Baja British Columbia. Somewhere beneath us lay the last ammonite fossil. Soon another island began to loom large, a place called Lummi Island, composed entirely of 50-million-year-old sediments and fossils. The first piece of North America. Our boat became the true time machine, carrying us from older to younger rock, from lithic fragments of the Age of Dinosaurs to equally hard but more recently formed rocks, pieces left from the earliest times of the Age of Mammals. It carried us over large stretches of the planet as well: from a land that once sat off Mexico to a land that has been here, in this corner of North America, for an equally long time.

It was still a beautiful day when we arrived at the boat launch and readied the boat for its tow to Seattle, back to the customary routines of life. But I was already plotting. With this boat I can rapidly reach every small island in the region or simply sit offshore of new places, new research regions. And who knows, perhaps some day this boat will indeed take me to the Mesozoic mainland, where the dinosaurs still roam.

References

BANNON J, D BOTTJER, S LUND, AND L SAUL. 1989. Campanian Maastrichtian stage boundary in southern California: Resolution and implications for large-scale depositional patterns. *Geology* 17: 80–83.

BIRKELUND T, JM HANCOCK, MB HART, PF RAWSON, J REMANE, F ROBASZYNSKI, F SCHMID, AND F SURLYK. 1984. Cretaceous stage boundaries—proposals. *Bull Geol Soc Denmark* 33: 3–20.

BURNETT J, J HANCOCK, W KENNEDY, AND A LORD. 1992. Macrofossil, planktonic foraminiferal and nannofossil zonation at the Campanian Maastrichtian boundary. *Newslet Stratigraphy* 7: 157–172.

BUTLER RF, WR DICKINSON, AND GE GEHRELS. 1991. Paleomagnetism of coastal California and Baja California: Alternative to large-scale northward transport. *Tectonics* 10 (5): 978–985.

CHRISTENSEN WK. 1988. Upper Cretaceous belemnites of Europe: State of the art. In Streel, M., and M.J.M. Bless. *The Chalk District of the Euregio Meuse-Rhine* (pp. 1–16). Selected papers on the Upper Cretaceous. Naturhistorisch Museum, Maastricht, Liège University.

COBBAN W. 1994. Diversity and distribution of late Cretaceous ammonites in the Western Interior. In Caldwell and E. Kauffman, eds. *Evolution of the Western Interior Basin* Geol Assoc Canada, Special Paper 39: 435–451.

FILMER P AND J KIRSCHVINK. 1989. A paleomagnetic constraint on the Late Cretaceous paleoposition of northwestern Baja California, Mexico. *J Geophys Res* 94: 7332–7342.

FRY J, D BOTTJER, AND S LUND. 1985. Magnetostratigraphy of displaced upper Cretaceous strata in southern California. *Geology* 13: 648–651.

GALE A, P MONTGOMERY, W KENNEDY, J HANCOCK, J BURNETT, AND J MCARTHUR. 1995. Definition and global correlation of the Santonian-Campanian boundary. *Terra Nova* 7: 6112–622.

GRADSTEIN F, ET AL. 1994. A Mesozoic time scale. *J Geo Res* 99: 24,051–24,074.

HAGGART J. 1990. New and revised ammonites from the upper Cretaceous Nanaimo Group of British Columbia and Washington State. *Geol Surv Canada Bull* 396: 181–221.

HAGGART J, AND P WARD. 1984. Late Cretaceous stratigraphy of the northern Sacramento Valley, California. *GSA Bull*, 95: 618–627.

HANCOCK J, AND A GALE. 1996. The Campanian Stage. *Bull Inst Roy Sci Nat Belg* 66: 103–110.

HANCOCK J, AND W KENNEDY. 1993. The high Cretaceous ammonite fauna from Tercis, Landes, France. *Bull Inst Roy Sci Nat Belg* 63: 149–209.

———. 1988. Mesozoic and Cenozoic chronostratigraphy and cycles of sea level change. *Society of Economic Paleontologists and Mineralogists Special Publication* 42: 73–108.

HANCOCK J, N PEAKE, J BURNETT, A DHONDT, W KENNEDY, AND R STOKES. High Cretaceous stratigraphy at Tercis. *Bull Inst Roy Nat Bel* 63: 133–148.

HAQ B, J HARDENBOL, AND P VAIL. 1987. The chronology of fluctuating sea levels since the Triassic. *Science* 235: 1156–1166.

HARDENBOL J, J THIERRY, M FARLEY, T JACQUIN, P DE GRACIANSKY, AND P VAIL. Cretaceous chronostratigraphy. In P. De Graciansky, ed. *Sequence Stratigraphy in European Basins*. SEPM Special Publication. In press.

HICKS J. 1993. Chronostratigraphic analysis of the foreland basin sediments of the latest Cretaceous, Wyoming, USA. Yale University doctoral thesis.

HOWELL D. 1993. *Principles of Terrane Analysis*. London: Chapman and Hall.

KAUFFMAN E, AND R KESLING. 1960. An Upper Cretaceous ammonite bitten by a mosasaur. *University of Michigan Museum Paleontology Contribution* 15: 193–248.

KENNEDY WJ. 1986. Campanian and Maastrichtian ammonites from northern Aquitaine, France. *Palaeontological Association of London. Special Papers in Palaeontology*, 36: 145.

———. 1989. Thoughts on the evolution and extinction of Cretaceous ammonites. *Proceedings of the Geologists Association* 100: 251–279.

LILLEGRAVEN J. 1991. Stratigraphic placement of the Santonian-Campanian boundary in the North America Gulf Coastal Plain and Western Interior. *Cret Res* 12: 115–136.

MACLEOD N, ET AL. 1997. The Cretaceous-Tertiary biotic transition. *J Geol Soc London* 153: 265–292.

MARKS P. 1984. Proposal for the recognition of boundaries between Cretaceous stages by means of planktonic foraminiferal biostratigraphy. *Bull Geol Soc Denmark* 33: 163–169.

MCARTHUR J, W KENNEDY, M CHEN, M THIRLWALL, AND A GALE. 1994. Strontium isotope stratigraphy for Late Cretaceous time: Direct numerical calibration of the Sr isotope curve based on the U.S. Western Interior. *Paleo Paleo Paleo* 108: 95–119.

NELSON B, K MACLEOD, AND P WARD. 1991. Rapid change in strontium isotopic composition of sea water before the Cretaceous/Tertiary boundary. *Nature* 351, 644–647.

OBRADOVICH J. 1993. A Cretaceous time scale. In Caldwell and E. Kauffman, eds. *Evolution of the Western Interior Basin*. Geol Assoc Canada, Special Paper 39: 379–396.

PESSAGNO E. 1967. Upper Cretaceous planktonic foraminifera from the western gulf coastal plain. *Paleontological Americana,* 5: 243–442.

PREMOLI-SILVA I AND W SLITER. 1994. Cretaceous foraminiferal biostratigraphy and evolutionary trends from the Bottaccione section, Gubbio, Italy. *Pal Ital* 82: 1–89.

SAUL L. 1988. New Late Cretaceous and early Tertiary Perissityidae (Gastropoda) from the Pacific Slope of North America. *Cont Sc, LA County Museum,* 400: 25.

SLITER W. 1968. Upper Cretaceous foraminifera from southern California and northwestern Baja California, Mexico. *University of Kansas Paleontological Cont.* 49, 141.

———. 1984. Cretaceous foraminiferans from La Jolla, California. In P Abbot, ed. *Upper Cretaceous Depositional Systems, Southern California, Northern Baja California.* SEPM book 49, 17–24.

SUGARMAN P, K MILLER, D BUKRY, AND M FEIGENSON. 1995. Uppermost Campanian Maastrichtian strontium isotope biostratigraphic and sequence stratigraphic framework of New Jersey coastal plain. *GSA Bull.* 107: 19–37.

TOSHIMITSU S, T MATSUMOTO, M NODA, T NISHIDA, AND S MAIYA. 1995. Toward an integrated mega-, micro- and magnetostratigraphy of the Upper Cretaceous in Japan. *J Geol Soc Japan* 101: 19–29.

VEROSUB K, J HAGGART, AND P WARD. 1989. Magnetostratigraphy of Upper Cretaceous strata of the Sacramento Valley. *California Bull Geol Soc Am* 101: 521–533.

WARD, P. 1978a. Revisions to the stratigraphy and biochronology of the Upper Cretaceous Nanaimo Group, British Columbia and Washington State. *Can J Earth Sciences* 15: 405–423.

WARD P. 1978b. Baculitids from the Nanaimo Group, Vancouver Island and Washington State. *J Paleontol* 52: 1143–1154.

WARD P, AND J HAGGART. 1981. The Upper Cretaceous ammonite and inoceramid bivalve succession at Sand Creek, Colusa County, California, and its implications for establishment of an Upper Cretaceous Great Valley Sequence ammonite zonation. *Newslet Stratigraphy*, 10, 140–147.

WARD PD, AND WJ KENNEDY. 1993. Maastrichtian ammonites from the Biscay region (France and Spain). *J Paleontol*, Memoir 34.

WARD P, K VEROSUB, AND J HAGGART. 1983. Marine magnetic anomaly 33–34 identified in the Upper Cretaceous of the Great Valley Sequence of California. *Geology* 11: 90–93.

Index

Absolute temperature scale, 36
Accretion tectonics, 84
Age of Dinosaurs, 128–129
Allonautilus genus, 185–186
    evolutionary history of, 188
Alvarez, Walter, 54–56, 65
Ambiguity, ix
Ammonite, 42
Ammonite septa, 155
Ammonite sutures, 151–160
Ammonites, xv, 11, 23, 102, 112,
    171–172
    Baculites, 30–32

best-preserved, 42
Cretaceous, 18, 24
evolution of, 42–43
in Haslam Formation, 113–114
from Hornby Island, 74
in Late Cretaceous Period, 33–34
Mesozoic, 213, 217–221
mosasaurs "eating," 135–146
planispiral, 219
species of, 15
Tunisian, 122–123
uncoiled, 128
virtual, 147–168

Ancient environments, sea level change and, 113–119
Ancient giant clams, 192–195
Ancient predation, evidence of, 130–135
Anston, F. W., 40
Apatite, 148
Appalachian Mountains, 83
Argon, 39
Ash, volcanic, 42
Asia and India, collision of, 82
Asteroid, 56
Athenaeum faculty club, 99
Atomic weight, 39
Australia, Indonesia and, 85
Autotrophs, 195–196

Baculites fossils, 30–32
Baculites inornatus, 69
Baja British Columbia, 73–104
Baja British Columbia hypothesis, 97
Baja California, 77, 99–102
Bakker, Bob, 140–142
Barnes Island, 28
Basalt, Mesozoic-aged, 80
Bay of Fundy, 115
Beard, Graham, 129
Beck, Merle, 91, 98
Becquerel, A., 38
Bentonite, 41
Biological structures, 148
Biomass, 196
Biostratigraphy, 28–32
Bipolar magnets, 47
Bostrychoceras polyplocum ammonite, 123
Bottjer, David, 64
Brandon, Mark, 97
Breaks in fossil shells, 130

Bridgewater Treatise (Buckland), 157
British Columbia, Baja, 73–104
Brunhes Normal Epoch, 51
Buckland, Dean, 156–157
Burgess Shale creatures, 150

California, 89–91
    Baja, 77, 99–102
Cambrian Period, 17
Cambrian System, 17
Campanian Stage, 21
Canada
    Geological Survey of, 29
    Western, morphogeological belts of, 92
Canadian fossils, 30
Canadian Gulf Islands, 191–192
Carbon-14, 39
Cascade volcanoes, 89
Catastrophes, global, 17
Catastrophic global climate change, 120
Catastrophism, 21
Cenoceras, 182, 186
Cenozoic Era, 11, 17
Cephalopods, 149
    fossil, 156
Chain saw as coring device, 57, 59
Chambered nautilus, 173
Chemoautotrophic bacteria, 202
Chico Creek, 60–64
Chicxulub comet impact, 74, 122, 172
Chromatophores, 218
Cladistics, 179
Clams
    Cretaceous, 212–213
    giant, 33
        ancient, 192–195
    inoceramids, xv, 112, 193–204

Clamshells, 8
Classification of life, 170–175
Climate change, catastrophic
    global, 120
Clocks
    magnetic, 45–69
    radiometric, 33–43
Cobbles, 109
Coelacanth, 181
Cold methane seeps, 201–204
Comet impact, Chicxulub, 74, 122, 172
Communities, 195
    hydrothermal vent, 200–202
Compaction shallowing, 104
Composite material, 143
Concrete, 143
Concretions, xiv–xv
Conglomerates, 109–110
Coniacian Stage, 19, 21
Contact, zonal, 26
Continental drift, 73, 76–89
Continental glaciation, 116
Convection cells, 81
    lithic, 78
Coring device, chain saw as, 57, 59
Courtney museum, 128
Cowan, Darrel, 97
Craton, travel paths of Western North
    America relative to, 93
Cretaceous ammonites, 18, 24
Cretaceous clams, 212–213
Cretaceous Period
    dates of, 61
    end of, x
Cretaceous stages, magnetochrono-
    stratigraphy and, 68
Cretaceous System, 17, 19, 25

Cretaceous/Tertiary boundary, see "K/T"
    boundary
Cretaceous time, 43
Cross beds, 110
Curie, Marie, 38
Curie point, 63

Daniel, Sophie, 141
Daniel, Tom, 141–144,
    162–168, 223
Darwin, Charles, 35, 36, 171
Data, interpretation of, x
Dating
    magnetostratigraphic, 55
    radiometric, 35
Decay, radioactive, 38–41
Declination of earth's magnetic
    field, 48
Deep sea cores, 54
Deep Sea Drilling Program, 57
Denton, Eric, 160
Derived characters, 179
Deserts, 106
Devonian Period, 161–162
D'Halloy, D'Omalius, 19
Dinofest 2, 140–142
Dinosaurs, xv
    Age of, 128–129
Direction
    paleomagnetic, 50
    sense of, 100
DNA, 170
D'Orbigny, Alcide, 19, 21, 25, 27, 28,
    43, 54, 156
Downey, Roger, 128
Du Toit, Alexander, 78, 79–80
Dzik, J., 178

Earth
  age of, 5, 34–41
  earthquake waves through, 50
  iron core of, 48
  as magnet, 47–48
  magnetic field of, 48
  onion analogy for, 81
Earthquake waves through earth, 50
Ebb tides, 115
El Kef, Tunisia, 121–122
Elasmosaur, 128–129
Environments, ancient, sea level
  change and, 113–119
Eocene-aged rocks, 96
Eustatic sea level change, 117
Evolution, 148–149
  theory of, 35, 36
Extension Formation, 115
Extinction controversy, 56
Extinctions, mass, 17, 56, 106, 119

Faculty clubs, 99
FEA (finite element analysis), 164
Feldspar, 42
Finite element analysis (FEA), 164
Flood tides, 115
Flower, Rousseau, 156
Food chains, 195
Foraminiferans, 25, 199
  planktonic, 55
Fossil Bay, xiii, xv
Fossil Beach, 75, 129
Fossil-bearing rocks, viii
Fossil cephalopods, 156
Fossil prizes, 14
Fossil shells, breaks in, 130
Fossiliferous outcrops, xiv

Fossils, xv, 7, 146
  Baculites, 30–32
  Canadian, 30
  law of superposition of, 12
  living, 175–178
  from measured sections, 13
  as relative indicators of time, 16
  in sedimentary rocks, 32
Functional morphology, 147, 150

Gabb, William, 22–23, 25, 27, 32, 56,
  60–61
Gabriola Island, 192–193
Garver, John, 97
Gauss Normal Epoch, 51
Gel electrophoretics, 184
Genetic drift, 189
Geological maps, 16
Geological Survey of Canada, 29
Geological time scale, 5
  historical development of, 20
Geology, 4, 86, 87
Geomagnetic polarity time scale
  (GPTS), 46–53
Geomagnetic poles, 80
Geophysics, 86–87
Georgia Strait, 127
Giant clams, ancient, 192–195
Gilbert Reversed Epoch, 51
Gills, 148
Gilpin-Brown, John, 160
Glacial tillites, 80
Glaciation, continental, 116
Global catastrophes, 17
Global climate change, catastrophic, 120
Gold Rush, 60
Gondwana Sequence, 80

Gondwana succession, 78–79

Gondwanaland, 79–80

Gould, Stephen Jay, 7, 150, 159, 187

GPTS (geomagnetic polarity time
    scale), 46–53

Granite, 34

Gulf Islands, 113

Haggart, Jim, 45, 64

Half-life, 39

Hardness of rocks, 6

Haslam Formation, 113–114
    ammonites in, 113–114

Heat flow, 36

Hell's Canyon, 96

Helmuth, Brian, 165

Hewitt, Roger, 161

High stands, 118

Himalayas, 82

Historical development of geological
    time scale, 20

Hooke, Robert, 154–156

Hoplitoplacenticeras, 69

Hoppe, Kathryn, 200

Hornby Island, xvi, 74–76
    ammonites from, 74

Horner, Jack, 140

Hurtado, Jose, 102, 103

Huxley, Thomas, 37

Hydrostatic pressure, 153

Hydrothermal vent communities,
    200–202

Ice Ages, 8, 116

Ichthyosaurs, 131, 132

Igneous rocks, 41

Iguanas, marine, 133

Implosion, 152, 153

Inclination of earth's magnetic field, 48

Inclination shallowing, 104

India and Asia, collision of, 82

Indonesia, Australia and, 85

Inoceramids, xv, 112, 193–204

Insular Superterrane, 97

Intensity of earth's magnetic field, 48

Intermontane region, 97

Interpretation of data, x

Intertidal zonations, 10

Iridium, 56

Iron core of earth, 48

Irving, Ted, 96–98

Isotopes, 34, 39, 198

Isotopic pairs, 39

Isotopic ratios, oxygen, 198–200

Jeletzky, George, 29–32

Johnson, Paul, 139

Jurassic Park, 150

K/Ar (potassium/argon) method, 39–40

"K/T" boundary (Cretaceous/Tertiary
    boundary), 17, 66, 119, 121
    mass extinction at, 17, 56, 106, 119

Kase, Tatsuro, 139

Kauffman, Erle, 135–138, 140–143, 145

Kelvin, Lord (William Thomson),
    35–39, 87

Kennedy, W. James, 156

Kesling, A., 135–138

King, Clarence, 37

Kirschvink, Joseph, 99–104

Komodo dragon, 130–131, 220

Kullman, Jurgen, 159

Kummel, Bernard, 177–178

Landman, Neil, 186

Larson, Neil, 142

Late Cretaceous Period, North America in, 33–34

Late Cretaceous Stage, 21

Lehmann, Ulrich, 156

Life
classification of, 170–175
sea level change and, 119–121

Limestone, 16
shallow-water, 54

Limestone bodies, 202

Limpets, 134, 139, 146

Lithic convection cells, 78

Lithification, 8

Living fossils, 175–178

Low stands, 118

Lowrie, William, 54–55, 65

Lummi Island, 224

Lyons, Jerry, x

MacLeod, Ken, 200

Magma, 34

Magnet(s)
bipolar, 47
earth as, 47–48

Magnetic anomalies, formation of, 51

Magnetic clocks, 45–69

Magnetic field of earth, 48

Magnetic pole(s)
modern north, 49
secular movement of, 48

Magnetic resetting of mineral grains, 88

Magnetism, rock, 87–88

Magnetite, 53

Magnetochronostratigraphy, Cretaceous stages and, 68

Magnetometers, 58, 63, 103

Magnetostratigraphic dating, 55

Magnetostratigraphy, 69

Magnetozone 33R, 69

Mantle, 81

Maples, Chris, 166

Mapping of North America, 22

Maps, geological, 16

Marine iguanas, 133

Martin, Arthur, 138

Mass extinctions, 17, 56, 106, 119

Mass spectroscopes, 40–41

Matsumoto, Tatsuro, 31, 178

Matuyama Reversed Epoch, 51

McPhee, John, 84

Measured sections, fossils from, 13

Memories, ix

Mesozoic-aged basalt, 80

Mesozoic ammonites, 213, 217–221

Mesozoic Era, viii, 8–9, 11, 17
Sucia Island in, science fiction of, 208–222

Methane seeps, cold, 201–204

Meyerhoff, Al, 77

Microfossils, 55

Microplates, 84–85

Mid-ocean ridge spreading center, 51

Miller, A. K., 176–178, 187

Mineral grains, magnetic resetting of, 88

Mining, 5–6

Modern north magnetic pole, 49

Molluscan shells, 143

Monitor lizards, 131

Morphogeological belts of Western Canada, 92

Morphology, functional, 147, 150

Mosasaurs, xvi, 129, 131–147, 219–220
   "eating" ammonites, 135–146
Mount, Jeff, 65–66
Mt. Baker, 91
Mt. Stuart, 91–94, 98
Mountains
   complexity of, 84
   formation of, 82
   primitive, 6
   secondary, 6
Mudstones, 74
Museums, 28
   natural history, 28–29

Nanaimo Group, xvii, 98, 113
Nanaimo Group strata, locations of, xviii
Natural history museums, 28–29
Naturalists, 35
Nautiloids, fossil, xv
Nautilus, 138, 167, 171
   ancestry of, 169–189
   chambered, 173
   classifying, 180–185
   pearly, 214–215
Nautilus belauensis, 180, 184
Nautilus genus, 174
   evolutionary history of, 188
Nautilus macromphalus, 180, 184, 189
Nautilus pompilius, 180, 182–184, 189
Nautilus scrobiculatus, 180–185
Nautilus shells, 141
Nautilus stenomphalus, 180, 184
Neopilina, 181
New Caledonian barrier reefs, 176
Newton, Isaac, 154

North America
   growth of, 83
   in Late Cretaceous Period, 33–34
   mapping of, 22
   Western, *see* Western North America
North magnetic pole, modern, 49
Nuclear submarine *Thresher*, 152

Ocean basins, 54, 77–78
Ocean Drilling Program, 54
Oceanic crust, 52
Oceans, temperature of, 198–204
Octopuses, 149
Offlap, 118
Onion analogy for earth, 81
Onlap, 118
Oppel, Albert, 158
Orcas Island, xii–xiii, 7, 12, 28
Oxygen isotopic ratios, 198–200

Pacific Plate, descending, 85
Paleoecology, 196
Paleomagnetic direction, 50
Paleomagnetic time machine, 65
Paleomagnetics, 62, 86–88
Paleomagnetism, 46–52, 75
   terrane analysis using, 91
Paleontology, viii
Paleozoic Era, 17
   start of, 16
Past, ix
Pearly nautiluses, 214–215
Permian Period, 162
Permian System, 17
Perpetual motion machines, 36
Pfaff, A., 158

Phenetics, 178–179
Phillips, John, 17
Phylogenetic systematics, 174
Physicists, 208
Pinna clam, 111
Placenticeras ammonites, 139
Planispiral ammonites, 219
Plankton, 197
Planktonic foraminiferans, 55
Plate tectonic theory, 52
Plate tectonics, 76–89
Plates, 81–82
Plesiosaurs, xvi, 131, 132
Polarity, reversals of, 48–52
Polarity Epochs, 51
Polarity time scale, 52
Poles, geomagnetic, 80
Potassium/argon (K/Ar) method, 39–40
Predation, ancient, evidence of, 130–135
Primitive mountains, 6
Pterosaurs, 210
Puget Sound, xix, 10
Puntledge River, 128–129

Queen Charlotte Islands, 97
Quendstedt, Albert, 158

Radioactive decay, 38–41
Radioactivity, 34, 38–39
Radiometric clocks, 33–43
Radiometric dating, 35
References, 225–229
Regression, 118
Reheated rocks, 63, 98
Relative time, 4

Reversals of polarity, 48–52
Rifting process, 86
Rock magnetism, 87–88
Rocks, viii, xi
  age of, 3–7
  Eocene-aged, 96
  fossil-bearing, viii
  hardness of, 6
  igneous, 41
  reheated, 63, 98
  sedimentary, *see* Sedimentary rocks
  Vancouver Island, 101
  volume of, within spreading centers, 117
Rocky Mountains, 34
Roentgen, Wilhelm, 38
Roux, Erica, 143–144
Royal Tyrell Museum, 138–139
Rutherford, Ernest, 38

Salt clock, 37
San Andreas fault, 82–83
San Francisco, 89
San Juan Island, xii, 10
Sandstone, 110–111
Saunders, Bruce, 161, 165, 180–186
Scallops, 10
Schindewolf, Otto, 159
Science fiction of Sucia Island in Mesozoic Era, 208–222
Science magazine, 103
Scientific method, vii
Sea level, 115
Sea level change, 105–106, 116–118
  ancient environments and, 113–119
  eustatic, 117
  life and, 119–121

Sea level curve, worldwide, 124
Sea shells, 130
Sea turtles, 138
Seattle, 99
Second law of thermodynamics, 36
Secondary mountains, 6
Secular movement of magnetic
    poles, 48
Sedgewick, Adam, 16–17
Sedimentary beds, 9
Sedimentary rocks, viii, 105–111
    accumulation of, 106
    fossils in, 32
    of Sucia Island, 107–112
    thickness of, 37
Sedimentary structures, 106
Sedimentology, 106
Seilacher, Dolph, 159
Seismic stratigraphy, 118
Sense of direction, 100
Septa
    ammonite, 155
    internal, 154
Shale, 114
    Nanaimo Group, xvii
Shallow-water limestone, 54
Shimansky, V., 178
Sierra Nevada Mountains, 23
Silicon Graphics workstations,
    163–164
Simpson, George Gaylord, 159
Skagit Valley, xii
Smit, Jan, 67
Smith, William, 15–16, 19, 23, 26,
    28, 156
Soddy, Frederick, 38
Sollas, William, 37

South Africa, 79
Spreading centers, 52, 80
    mid-ocean ridge, 51
    volume of rock within, 117
Squires, Richard, 186–187
Stages as units of time, 19, 21
Stratigraphy, 9, 11–12, 78
    seismic, 118
Stratotypes, 64
Strike slip motion, 89
Subduction zones, 78, 81
Submersibles, 152
Sucia Island, x, xii–xix, 7, 98
    age of, 3–7
    in Mesozoic Era, science fiction of,
        208–222
    sedimentary rocks of, 107–112
    site of, xv–xix
    stratigraphic location of, 25–28
    telephone book analogy of, 107–109
    voyage to, xii–xiii
    winter trip to, 12–14
Suess, Eduard, 79
Superposition of fossils, law of, 12
Suspect terrane, 73, 84–86
Sutures, ammonite, 151–160
Systematics, 170
    phylogenetic, 174

Taxon/taxa, 14, 178
Taxonomy, 170
Tectonics, accretion, 84
Teichert, C., 178
Telephone book analogy of Sucia Is-
    land, 107–109
Temperature of oceans, 198–204
Temperature scale, absolute, 36

Terrane, 86
  suspect, 73, 84–86
  Wrangellia, 94–97
Terrane analysis using paleomagne-
    tism, 91
Tertiary mountains, 6
Tessier, Robert, 91
Texada Island, 102–103
Thermodynamics, second law of, 36
Thompson, J. J., 40
Thomson, William (Lord Kelvin),
    35–39, 87
Thought experiments, 208
Thresher nuclear submarine, 152
Tides, 115
Tillites, glacial, 80
Tilting, 94
Time
  Cretaceous, 43
  fossils as relative indicators
      of, 16
  relative, 4
  standard for, 43
Time-keeping, 4–5
The Time Machine (Wells), x
Time machines, vii
  detailed observations as, 124
  devices as, ix
  functional morphology as, 147
  informed imagination as, 222
  microcomputers as, 168
  natural history museums as, 29
  one of first, 16
  paleomagnetic, 65
  radiometric methodology as, 34
  from science, xvii–xviii
  scientific method as, 146

Time scale
  geological, *see* Geological time scale
  geomagnetic polarity (GPTS),
      46–53
  polarity, 52
Transgression, 118
Trask, Mike, 128–129
Trigonia clam, 111
Trigoniids, xv
Trophic group analysis, 195–197
Tunisia, 120–124
Tunisian ammonites, 122–123
Tunnicliffe, Verena, 203–204

Umhoefer, Paul, 97
Underwater volcanoes, 52
Uniformitarianism, 106
  principle of, 11
Urey, Harold, 198

Van Andel, Tjeerd, 200–201
Vancouver Island, xvi, 23, 127
Vancouver Island region, 73
Vancouver Island rocks, 101
Vent fauna, 201
Verosub, Ken, 57, 59, 62, 76
Virtual ammonites, 147–168
Vital effects, 200
Volcanic ash, 42
Volcanoes, 34
  Cascade, 89
  underwater, 52

Wainright, Steven, 149, 162
Waldron Island, 27
Wegener, Alfred, 78, 79–80
Wells, H. G., x

Wells, Martin, 149
Westermann, Gerd E. G., 153,
    159–161, 166
Western Canada, morphogeological
    belts of, 92
Western Interior Seaway, 33, 42, 135
Western North America, 90
    travel paths of, 93
Wiedmann, Jost, 159, 178

Willey, Arthur, 181, 182
Woodruff, David, 184
Worldwide sea level curve, 124
Wrangellia terrane, 94–97
Wray, Charles, 184

Zonal contact, 26
Zonations, intertidal, 10
Zumaya, Spain, 65–67